VOICES

OF THE

FOOD

REVOLUTION

VOICES
OF THE
FOOD
REVOLUTION

YOU CAN HEAL YOUR BODY AND
YOUR WORLD—WITH FOOD!

JOHN
ROBBINS

AND

OCEAN
ROBBINS

Conari Press

First published in 2013 by Conari Press, an imprint of
Red Wheel/Weiser, LLC
With offices at:
665 Third Street, Suite 400
San Francisco, CA 94107
www.redwheelweiser.com

Library of Congress Cataloging-in-Publication Data
Robbins, John.
  Voices of the food revolution : you can heal your body and your
world—with food! / John Robbins and Ocean Robbins.
       pages cm
  Summary: "Join John and Ocean Robbins for 21 intimate, game-
changing conversations with some of the world's leading "food
revolutionaries": scientists, doctors, teachers, farmers, economists,
activists, and nutritionists working on food issues today"—Provided
by publisher.
  ISBN 978-1-57324-624-8 (pbk.)
  1. Diet in disease. 2. Diet therapy. 3. Nutrition. I. Robbins, Ocean,
1973- II. Title.

RM216.R62 2013
615.8'54--dc23

                           2013006185

Cover design by Stewart Williams
Interior by Maureen Forys, Happenstance Type-O-Rama

Printed in the United States of America
MAL

10 9 8 7 6 5 4 3 2 1

The paper used in this publication meets the minimum requirements
of the American National Standard for Information Sciences—Perma-
nence of Paper for Printed Library Materials Z39.48-1992 (R1997).

Dedicated to the day, hopefully soon, when no child, anywhere on earth, lacks healthy food or abundant love.

Contents

Acknowledgments

We want to express our deepest gratitude to our wives, Deo Robbins and Michele Robbins, who have put up with us through thick and thin for forty-six years and nineteen years respectively, and provided us with abundant wisdom, boundless support, and infinite love.

Thank you also to the other members of our family and support team. To Veronica Monet, for the countless ways she has nurtured and encouraged us. To Bodhi and River Robbins, our twins/grandtwins, for teaching us about unconditional love and the power of play. And to Tom Callanan, Elizabeth Hendren, and Megan Saunders for all the ways their love and labor have enriched our lives and literally made this book possible.

Thank you to the brilliant and inspiring visionaries we were blessed to interview as we developed this book: Bill McKibben, Caldwell Esselstyn, M.D., Dean Ornish, M.D., Elizabeth & Dennis Kucinich, Frances Moore Lappé, Gene Baur, Geneen Roth, Jeffrey Smith, Joel Fuhrman, M.D., Joseph Mercola, D.O., Kathy Freston, Marianne Williamson, Michele Simon, Morgan Spurlock, Neal Barnard, M.D., Nicolette Niman, Nikki Henderson, Raj Patel, Ph.D., Ronnie Cummins, Rory Freedman, T. Colin Campbell, Ph.D., and Vandana Shiva, Ph.D. We are profoundly grateful to each and every one of you for your beautiful and courageous work, for your eloquence, and for the generosity with which you share your wisdom.

Thank you to Lionel Peter Church, Gregg Boggs, Kaia Van Zandt, Candice Csaky, and Brandon Jennings, who did so much infrastructural work to make this possible.

Thank you to all of the affiliates and partners who have made the launch of the Food Revolution Network such a success. This book might never have happened without you. In particular, thank you to Bonny Meyer, Chris Kaatz, Dawn Moncrief and A Well-Fed World, Eron Zehavi, Jeff Nelson, Linda Riebel, Sylvia Bass, Tamara West, and Tera Warner for your unique and vital contributions.

We also want to thank Caroline Pincus and the whole team at Conari. It is always a privilege and a pleasure to work with you.

We want to thank all of the farmers, farm workers, retailers, community leaders, chefs, educators, advocates, conscious consumers, and community leaders who are working for more nutritious, local, organic, natural, conscious, and sustainable food.

And we want to thank you, dear reader, for every step you take to live the food revolution. You inspire us, and you are helping to change this world for the better.

~~~~~ Introduction ~~~~~

Thank you for participating in one of the most important conversations about our food that has ever taken place.

Our food chain is in crisis. Big agribusiness has made profits more important than your health—more important than the environment—more important than your right to know how your food is produced.

But beneath the surface, a revolution is growing.

From rural farms to urban dinner plates, from grocery store shelves to state ballot boxes, people are rising up and taking action. We're reclaiming our food systems and our menus, and we're taking responsibility for our health.

Today there's a huge and growing demand for food that is organic, sustainable, fair trade, non-GMO, humane, and healthy. In cities around the world, we're seeing more and more farmers' markets, and more young people getting back into farming. Grocery stores (even big national chains) are displaying local, natural, and organic foods with pride. The movements for healthy food are growing fast, and starting to become a political force.

The days of skyrocketing obesity and chronic illness . . .

The days of families, companies, and governments being driven to bankruptcy by mounting health-care costs . . .

The days of unlabeled, genetically engineered foods spreading rampantly through our food system while family farmers are driven out of business . . .

Those days may be numbered.

Medical science today knows a lot about the impact our food choices have on our health, and on the world around us. But so far, that information has been far too slow to spread. Despite all the progress, hundreds of millions of people just keep getting sicker, and our planet keeps getting more polluted.

Even our doctors often are ignorant of how powerfully food impacts health. On average, physicians still receive only 24 hours of nutritional education in all their years of medical school. No wonder doctors are more likely to tell you how you can undergo an invasive and dangerous $50,000 procedure than they are to tell you how you can simply and easily change your diet to prevent illness in the first place.

In the absence of educated physicians and an informed public, commercial interests try to sell us nutritional ideas, breakthrough gimmicks, and pseudoscientific diets that have little basis in reality. Most of us are inundated every day with supposed "facts" that are dangerously misleading, or just plain wrong.

My dad, bestselling author John Robbins, and I decided that it was time to distill the wisdom of the greatest international experts on food and diet and share their leading-edge insights. We set out to discover what you need to know that no one is telling you, including what you can do to get healthy, and how you can even help the environment through your choices.

One of the things that stands out is that—no big surprise here—our current food system is a mess.

Large-scale industrialized food production is wreaking havoc on our forests, topsoil, air, water, and climate. Farm animals are being treated with tremendous cruelty, and farm workers are often exploited. Genetically engineered

"Frankenfoods" are being released, without adequate testing, into the food supply on a vast scale. Meanwhile, people are eating more and more artificial food—and getting fatter and sicker. In fact, more people are chronically ill today than at any time in the history of the world.

You probably already know that large-scale industrial agribusiness is controlling an expanding share of the world's food supply. They have huge advertising budgets to market highly processed, genetically engineered, chemical-laden, pesticide-contaminated pseudofoods. With all their lobbyists and political donations, they pretty well run the show in most of our government regulatory and agricultural agencies.

These corporations want to fill your plate with chemicals, and they're spending billions marketing processed foods that get people addicted for life. They'd like you to keep your mind closed and your voice silent. They'd like you, and all the rest of us, to keep eating foods that are unhealthy, because by eating these foods, we provide big profits for companies like McDonald's, Monsanto, and Coca-Cola. They'd like to keep you subordinate to their agenda, and so distracted that you won't raise a peep of protest.

Fortunately, you don't have to do that.

You already know that there's a huge link between your food and your health.

You know that a majority of the medical costs that are bankrupting families, companies, and nations could be eliminated with better nutrition.

But do you know the alternative? Do you know what the experts have found out about how to promote optimal health, and how to contribute to building a healthier society?

Now is the time to find out how you and your family can get informed and take action.

Have you ever been chronically ill? Worried about the health or survival of a loved one? Would you like to know how to lose weight, clean up your arteries, or defend yourself against cancer?

Do you sometimes wish you had more energy, got sick less often, and felt more confident about what to eat for optimal health?

For most of us, a good diet is the best gift we can give to ourselves and our loved ones... because it's the gift of lasting health. And it's not rocket science. Using simple and easy-to-remember steps, you can dramatically increase your chances of living a long and vibrant life.

Do you care about the world around you? Do you want your food choices to contribute to building a more sustainable and compassionate world?

You have a right to know the truth about what you eat, where it comes from, and what its impact is on your life and on the planet. The more you know, the more power you have to take meaningful action. The more you know, the better able you are to bring your food choices into alignment with your purpose and your passion. Your mind will be clearer, your heart will be more at peace, and your body will thank you for the rest of your life.

Right now there are many people concerned about the future of food who want reliable, up-to-date information from sources they can trust. We created this book as a way to help. Sure, there have been thousands of books, conferences, documentaries, and seminars on food. But we have hand-picked these leaders and experts to bring you the best of the best—experts who aren't beholden to commercial or political agendas, and who have made it their job to accurately find and effectively communicate the truth.

There's one other thing about this book that's pretty unique. The interviews are conducted by a world-renowned hero of the food movement. If you'll indulge me for a moment, I'd like to tell you a little family history.

My grandfather, Irvine Robbins, founded the Baskin-Robbins ice cream company, with all of its legendary 31 Flavors. My dad, John Robbins, grew up with an ice-cream-cone-shaped swimming pool. From his earliest childhood, he was groomed to join in running what became the world's largest ice cream company. That's right, you're reading about the food revolution from a family that actually has roots in the mass-marketing of frozen, sugar-laden butterfat.

But my dad walked away from all that. He said nope, I don't want to spend my life selling ice cream, thank you very much. He left behind the company, and with it any access to or dependence on the family wealth. He followed his own "rocky road," and wound up moving with my mom to a little island off the coast of Canada. There they lived very simply, grew their own food, practiced yoga and meditation, and had a kid they named "Ocean." That's me.

In 1987, John Robbins, the could-have-been-but-decided-not-to-be ice cream heir became the "rebel without a cone," when he inspired millions of people through publication of his groundbreaking, landmark bestseller, *Diet for a New America*.

For nearly three decades, my dad (and now my colleague), John Robbins, has been speaking out for healthy, sustainable, humane, and delicious food. He's gone from sometimes seeming like a voice in the wilderness, to being one of the world's leading authorities on the subject. His books have sold more than two million copies and have been translated into twenty-seven languages. He's keynoted hundreds of conferences and received numerous awards and accolades. Most important, his work has inspired countless doctors, scientists, farmers, activists, and everyday folks to find out the truth about the relationship between the food on our plate and the lives we create.

In this book, for the first time ever, one of the founders of the modern food movement engages in breakthrough dialogues with twenty-one of today's top food revolutionary experts.

These conversations are special. They utilize a combination of interviews and dialogues to capture decades of hard-won wisdom, and to bring it to you in an easily digestible and highly useful form. You may have read interviews before, but how often do you get top-notch experts engaging with a game-changing movement leader—all for your benefit?

I've had the pleasure of curating and editing this remarkable collection. It's been a privilege to learn from some of the most important voices of our time, and to help bring their crucial message to you.

With this book, we offer you a diverse, gourmet, tasty, and nutrient-rich powerhouse that's designed to help you move from being a medical time bomb to a health superstar, and from a frustrated spectator to an empowered agent of change.

I love food. I love eating it, I love preparing it, and I love sharing it with other people. Throughout the world, "breaking bread" together, or sharing a meal together, is an act of connection. Food bonds us to the world, to culture, and to one another.

As you dive into this book, our hope is that you find it as delicious as it is nourishing, and as useful as it is inspiring.

Thanks for joining us.

Bon appétit.

—*Ocean Robbins, spring 2013*

PART I

What Is the Optimal Diet for Human Beings?

The United States has the world's highest rates of obesity and chronic illness. People are spending more and more of their lives sick.

Health-care spending, which should more accurately be called disease-care spending, now consumes nearly 18 percent of the U.S. GDP, and it keeps rising. Three-quarters of this money is going to treatment of chronic diseases, most of which are preventable and linked to the food we eat.

The good news is that modern research gives us tremendous knowledge about the link between diet and disease, and about the real sources of health. Science knows, unequivocally, what it takes to dramatically increase the likelihood of living a thriving and vibrant life.

While most doctors receive less than 24 hours of nutritional education in 4 years of medical school, some have bucked the status quo and devoted decades of their lives to cultivating the wisdom their peers are so lacking.

We sought out the wisdom of some of the most seasoned experts—people whose programs are rooted in science, and whose results are unassailable. If you want the honest truth about your diet and your health, read on . . .

1

Dean Ornish, M.D.

Simple and Proven Breakthroughs That Are Changing the World

Dean Ornish, M.D., is one of the greatest medical pioneers in the world. His research has demonstrated—for the first time—that integrative changes in diet and lifestyle can reverse heart disease, turn on health-promoting genes, slow aging, and slow or even reverse early stage prostate cancer. Medicare and many of the largest insurance companies have made his program the first lifestyle-based approach they have ever covered. Chosen by Forbes *as "one of the seven most powerful teachers in the world," Dr. Ornish's work is changing the face of medicine.*

Dr. Ornish challenges the myth that you have to choose between what's good for you and what's fun for you. His research proves that better diet can lead to better sex, more energy, and a happier life.

JOHN ROBBINS: Your results have been extraordinary. Patients in your program see their angina reverse or decrease as early as the first few weeks. Blood flow to the heart improves, often in a month or less. After a year, even severely blocked coronary arteries become measurably less blocked. It seems that as the years go by there is even more reversal and more improvement. Have you seen comparative improvements in patients following the more moderate American Heart Association Guidelines or any other program that you know of?

DR. DEAN ORNISH: Moderate changes may be enough to prevent heart disease in some people, but they are usually not enough to reverse it. We were able to show in a scientific way that most traditional recommendations didn't go far enough. The more you change your diet and lifestyle, the more you improve in virtually every way we can measure, whether it is your heart disease improving, your PSA coming down, or your gene expression changing. We found that over 500 genes were changed in just three months, with the up-regulating or turning on of genes that prevent disease, and down-regulating or turning off of genes that help promote disease. Particularly what are called the RAS oncogenes that promote cancers of the prostate, breast, and colon were down-regulated. These processes are much more dynamic than anyone had realized. The more we look, the more we find.

Some people think taking a pill is easy and everyone will do it, but that changing diet and lifestyle is difficult, if not impossible, and hardly anyone will do it. What we're finding is actually the opposite. Adherence to most medications, whether they are cholesterol-lowering drugs or blood pressure pills, is only about 30 percent at three or four months. But we are getting 85–90 percent adherence after a year in Nebraska, Pennsylvania, and West Virginia, even though West Virginia leads the United States in heart disease. The reason is that the pill may not make you feel better, but changing your diet and lifestyle will. The better you feel, the more you want to keep doing it so you get into a virtuous cycle. That is one of the reasons people are continuing to do it—not just to live longer, but to live better.

JOHN ROBBINS: So they enter the program because they have some medical problem that they want to alleviate and they end up changing their life in a way that creates benefits across the board. It seems as though you have found an entry point into people's lives that is healing in a profound sense.

DR. DEAN ORNISH: Well it is, and that is why I love doing this work. You know, we are all going to die of something. The mortality rate is still 100 percent, it is one per person. But the question is not just how long we live, but also how well we live.

There is a belief that you have to choose between what is good for you and what is fun for you. But we're saying you can have both. You can have more fun, have better sex, sleep deeper, enjoy your food more, and not have all those aches and pains.

The ancient swamis, rabbis, priests, monks, and nuns didn't develop techniques like meditation, yoga, and so on to unclog their arteries or lower their blood pressure. But it turns out that they developed some really powerful tools for transformation that are also physiologically healing. I can't tell you how many patients have said things to me like, "Even if I knew I wouldn't live another day longer, I would still make these changes now that I know what they are like because my life is transformed."

What is most meaningful to me is how we can work with people to use the experience of suffering in whatever way they are feeling it as a catalyst and a doorway for transforming their lives. For some, it is physical suffering, because they have angina or chest pain. For others, it is the suffering of depression or isolation. If we can work at that level, then we find that people are much more likely to make lifestyle choices that are life enhancing rather than ones that are self-destructive.

To me it is really about transformation, and then on a physical level, just about everything we measure tends to get better. We're seeing dramatic improvements in things that were never measured before. We found that even telomerase increased by 30 percent in the first three months.

JOHN ROBBINS: Why it that important?

DR. DEAN ORNISH: Dr. Elizabeth Blackburn won the Nobel Prize in 2009 for her co-discovery of telomerase. Telomerase repairs and lengthens our telomeres, which are the ends of our chromosomes. Telomeres control aging, which in turn controls how long we live. So as your telomeres get shorter, your life gets shorter.

She had done a pioneering study with women who were caregivers of parents with Alzheimer's or kids with autism. The more stress the women reported feeling and the longer they reported feeling that way, the lower their telomerase and the shorter their telomeres were. It made headlines because it was the first study showing that even at the genetic level, chronic stress can actually shorten your life.

What was one of the more interesting findings of this study is that stress is not simply what happens to you, it is how you deal with it. You could have two women who were in very comparable life situations, but one was coping much better than the other. It wasn't the objective measure of stress that determined its effects on telomerase; it was the women's perception of it. So when people learn how to meditate and do yoga and use other methods to manage stress, they can be in the same job or the same family or the same environment and react in different ways.

JOHN ROBBINS: I am hearing two things. I am hearing that your research is showing that the kind of lifestyle changes you recommend actually affect gene expression, turning on disease-preventing genes and turning off genes that promote cancer, heart disease, and other diseases. I am also hearing that there is evidence that these same lifestyle changes have an affect on aging, because they actually impact the telomerase.

DR. DEAN ORNISH: Yes, we found that the telomerase increased by almost 30 percent in just three months, and ours is the only intervention to date that has been shown to do that. We are now looking to see the effects on telomere lengths.

JOHN ROBBINS: That is astounding information and a major breakthrough.

DR. DEAN ORNISH: It is a radically simple idea that when you eat healthier, when you love more, when you manage stress better, when you exercise moderately, and when you don't smoke, you are happier and your life is more fun.

JOHN ROBBINS: You have become primarily known for your nutritional wisdom and research, but you have also written in great depth about the healing power of relationships and intimacy. We live in a culture where loneliness probably kills more people than cigarettes. What are you learning about the connection between relationships and health?

DR. DEAN ORNISH: We get to know each other really well in some of my studies. After a while I asked some of my patients, "Why do you smoke? Why do you overeat and drink too much and work too hard and abuse yourself? These behaviors seem so maladaptive to me."

They would say: "You don't get it. You don't have a clue. These are adaptive behaviors, because they help us get through the day."

One patient said, "I've got twenty friends in this package of cigarettes and they are always there for me."

So it is true that loneliness kills more people than cigarettes, but keep in mind that it is often loneliness that causes people to smoke in the first place. Or they use food to fill the void or alcohol to numb the pain, or they work all the time to distract themselves.

To me, the function of pain is to say, "Hey, listen up. Pay attention. You are doing something that is not in your best interest." Then we start to say, "Oh, I have a different choice. I can do this instead of that," and then it comes out of your own experience, not because some authority figure told you.

JOHN ROBBINS: I want to take the example of your friend, former President Bill Clinton. He has had serious health problems for a long time. In 2004 he underwent a quadruple bypass to restore blood flow to his heart. A few years later, there were more problems and he had two stents placed inside one of his coronary arteries that had once again become clogged. After that he made a decision that I think you had probably been encouraging him toward for some time. That decision seems to have transformed his life. He has lost more than twenty-five pounds. The last time I heard from him, he said he was feeling healthier than ever. He may have become the world's most famous vegan. What can you share about President Clinton's health journey and your involvement with it?

DR. DEAN ORNISH: Well I love President Clinton. I think he is an amazing human being. I began working with the Clintons in 1993 when someone arranged for Mrs. Clinton to meet with me. She was interested in the research that we were doing and after showing her the research she said, "Would you work with the chefs who cook for us?" And I said, "Excuse me?" She repeated it and I said, "Of course, I would be honored to." So I brought in some of the top chefs that we have worked with over the years. We went to the White House and we worked with the chefs who cooked for the President. We also worked with the chefs from Camp David and from Air Force One. It was a great privilege to be able to do that.

I also began working with the President on his own health. He did make some changes in the way he was eating and it did lead to some benefits, but when his bypass clogged up, there was a press conference in which it was announced that it was all in his genes and his diet and lifestyle had nothing to do with the bypass clogging up. So I communicated with him and said, "The friends that I value the most are the ones who tell me what I need to hear, not necessarily what I want to hear. What you need to know is that it is not all in your genes. I say this not to blame you, but to empower you, because if it were all in your genes, you would just be a victim. There is a lot you can do. Again, that is the opportunity. It is not a criticism in any way. It is out of great respect and love."

I sent President Clinton some of the research and a couple of books that I have written on the topic, and when we met a week or so later, he said that he had decided to do it. I was really happy that he did that because I care about him so much. I was also happy because I think, whatever your politics, when the President of the United States—especially one who was known earlier in his life for eating particularly unhealthfully—makes a choice to eat a lot healthier, that sets a great example for everyone. So, as you say, he may have become the world's most well-known vegan. Whatever other considerable good that he does in the world, he will have added to it substantially by showing that this is possible.

JOHN ROBBINS: I want to share one of my favorite quotes of yours. You said, "I don't understand why asking people to eat a well-balanced, vegetarian diet is considered radical, while it is medically conservative to cut people open and put them on powerful cholesterol-lowering drugs for the rest of their lives."

DR. DEAN ORNISH: We pay for things that are dangerous, invasive, expensive, and largely ineffective. And we have a hard time believing that simple choices in our lives that we make each day—like what we eat, how we respond to stress, how much love we give each other, and how much we exercise—can make such a powerful difference. But they really do. That is probably our unique contribution: we use these very high tech, expensive, state-of-the-art measures working with first-rate scientists to show how powerful these very simple, low-tech, and low-cost interventions can be. Three-quarters of the 2.8 trillion dollars that we spend each year on health care in the United States is for chronic diseases that can often be prevented or even reversed by simply making these same kinds of changes. In every one of our studies, the more people have changed their diet and lifestyle, the more they have improved.

It is not like there is one diet and lifestyle intervention for heart disease and a different one for prostate cancer and a different one for Type 2 Diabetes and a different one for changing your genes or making your telomeres longer. It is really just to the degree that you move in this direction, we've found there is a corresponding benefit. It is not all or nothing. Being a vegan is too much for some people, so we say, "Okay great. Just do what you can. What matters most is your overall way of eating and living. Foods aren't good or bad, but some are healthier for you than others." So we have categorized foods from the healthiest (group one) to the least healthful (group five), and groups two through four are intermediate. If you are eating mostly four and five, you are eating mostly unhealthy food.

If you feel like making any changes, of course that is up to you. You can say, "Okay, maybe I will eat a little less four and five and maybe more one through three." Then you do that and you see if you feel better. If you start to feel better, then maybe you want to make even bigger changes, but again it is coming out of your own experience.

The problem with going on a diet is that when you go on a diet, sooner or later you are likely to go off it. Once you go off it, you may feel like you are a failure. Be compassionate with yourself. Just say, "Directionally I am going to go more towards a plant-based diet because I want to feel better. So if I indulge myself one day, then I will eat healthier the next. If I don't exercise one day, I will do a little more the next. If I don't have time to meditate for an hour, I will do it for a minute." Just the consistency is more important than anything, and that way you can't fail. It just becomes a way of living in the world rather than just a diet that you go on and off.

JOHN ROBBINS: It reminds me of the old story, The Tortoise and the Hare. Slow and steady wins the race.

DR. DEAN ORNISH: That's true. And it's also true that sometimes people do really well when they make big changes all at once because they feel so much better so quickly, and the effect of their changes is highly visible. For others it is easier to go with slow and steady.

We say, "Here are the risks and the benefits, the costs and the side effects. This is your life, and your responsibility.

I am only here to support you and make sure that you have all of the information that you can use to make intelligent choices. I'm here to support whatever you choose to do, whether it is drugs, surgery, lifestyle, or a combination."

JOHN ROBBINS: You have been engaged in a dialogue with Medicare for a long time that is now bearing fruit. One would hope that if you do good science it is going to change medical practice. But if there isn't reimbursement, then it might not.

DR. DEAN ORNISH: I am grateful for Medicare. They are now covering "Dr. Ornish's Program for Reversing Heart Disease" as a named program in the clinics, hospitals, and physician's offices where we train and certify. Having seen what powerful changes diet and lifestyle can make over the past thirty-five years of studies, it is great to be able to make them available. Reimbursement is an important determinant of medical practice and even medical education.

I learned a painful lesson when we opened a number of sites before we had reimbursement to cover our program. Even though we got excellent clinical outcomes, some of those sites closed down. They didn't close because the treatment wasn't working. They closed because it wasn't reimbursable.

If you change reimbursement, then everything follows. Without that even a thousand studies may not be enough to really change things.

After sixteen years of reviewing our work internally and externally, Medicare agreed to cover it. Now we can make it available to people everywhere, because most other insurance companies are following their lead.

We trained ten hospitals in West Virginia which has been the number one state in the country for heart disease, and 44 hospitals and clinics elsewhere, with many more to come. We trained the St. Vincent de Paul Society homeless clinic in San Francisco in our Spectrum Integrative Medicine Program and they have treated more than 15,000 people through it in the past year and a half. We will be offering our reversing heart disease program there as well which Medicare will reimburse, and then we'll clone it to St. Vincent de Paul homeless clinics throughout the country.

We are in the process of training lots of different people in our program and providing a truly integrative paradigm that incorporates the best of drugs and surgery when they are effective, but also addresses the more fundamental causes of why we get sick.

JOHN ROBBINS: You have become known for advocating a low-fat diet, and yet as you have so often pointed out, low fat is not synonymous with healthy. I have seen studies that purport to measure low-fat diets to see if there are any benefits, but they define "low fat" as an intake barely less than the norm in our culture, and then don't see benefits. There is barely any reduction in fat consumption, so you see barely any results.

DR. DEAN ORNISH: Right, well not only is there barely any reduction in fat, but also they usually put in tons of sugar. I am so sorry that I somehow have gotten this reputation as the low-fat guy. I think it came out of the debates I was in with Dr. Atkins. To me that is the least interesting part of the work that we do.

JOHN ROBBINS: What would you say is the role of fat in an optimally healthy diet?

DR. DEAN ORNISH: It depends. Three or four grams a day of omega-3 fatty acids taken during pregnancy can increase a child's IQ. For adults, they can reduce your risk of sudden cardiac death by up to 80 percent, because they raise the threshold for ventricular fibrillation. They can reduce your risk of prostate or breast cancer. They can help reduce inflammation. They can lower triglycerides. I think that those are good things to have in your diet.

If you are vegan, you can get omega-3 fats from the plankton-based supplements, and if you are not, you can get them from fish oil. I recommend that just about everybody take three or four grams a day of fish oil or the equivalent.

There is a consensus that trans fats are harmful in a number of different ways. There is a lot of controversy about whether saturated fat is harmful or not. I think it is, with the possible exceptions of the saturated fat that you find in chocolate, which, although it is dense in calories, may actually be good for your heart.

Total fat consumption matters in a number of ways, one of which is that fat is the most dense form of calories. Fat has nine calories per gram and protein and carbs have only four. So if you are eating less fat, you are getting fewer calories without having to eat less food. It is the volume of the food that really seems to stimulate satiety more than the caloric density of the food. An easy way to lose weight is to just reduce the amount of fat in your diet, because you are going to be getting the same quantity of food, but it will be less dense in calories, so you are going to be eating fewer of them.

JOHN ROBBINS: How about coconut?

DR. DEAN ORNISH: Coconut is high in saturated fat. I think the jury is out on coconut. I haven't seen enough good evidence either way.

JOHN ROBBINS: Olive oil is frequently touted as a health food and the center of the Mediterranean Diet. At the same time, canola oil is getting a bad reputation in certain circles. What do you think of this controversy?

DR. DEAN ORNISH: One of the main reasons people think that olive oil is good for you is because of the Lyon Study that took place in France. People consuming what was termed a Mediterranean Diet had an 80 percent lower risk of heart attacks than those who didn't. But if you actually looked at what they were eating on this so-called Mediterranean Diet, they were reducing their intake of saturated fats—meat and butter and dairy and so on. And they were eating a diet that was high in canola oil. What makes canola oil beneficial are the omega-3 fatty acids that are high in canola oil but are very low in olive oil. That is what they were eating in the Lyon Study, and so the Mediterranean Diet that was so beneficial was high in canola oil, more than in olive oil.

Now olive oil does have antioxidants. It has some good things. The problem with olive oil is that, like all oils, it is so dense in calories. One tablespoon of olive oil has about fourteen grams of fat. People dip their bread in it and soak up large amounts of oil, thinking it is somehow going to be good for them and not realizing that they are getting a lot of calories.

JOHN ROBBINS: I want to ask you about alcohol. I have always seen you as somebody who wants people to enjoy their life.

DR. DEAN ORNISH: Absolutely. What is the point otherwise?

JOHN ROBBINS: For many people, alcohol is part of their joy in life and their connection with other people. Yet of course, in excess it is addictive and terribly destructive. We are seeing some evidence that moderate alcohol consumption has actual health benefits. What is your take on this?

DR. DEAN ORNISH: I think that you should do things you enjoy. I don't prescribe alcohol. I don't proscribe alcohol. I say if you are going to drink, the studies show that it is best to keep it on average under two drinks a day. Two drinks means two glasses of wine, two cans of beer, two shots of whiskey, or the equivalent.

The studies show that people who drink moderately can derive some benefit. But these are people who like to hang out with their friends after work. They go to a bar or go to a restaurant for happy hour. It's a great place for people to get social support. So it's hard to separate how much of that benefit is due to the social support, and how much is due to moderate alcohol consumption. Rather than trying to sort out the relative parts, I would say they are probably both important. But I wouldn't tell somebody who is not drinking to start. I think that it is important that people feel like they have lots of ways of increasing their social support or managing stress. Some people may sit around drinking, and other people will do it in other ways.

JOHN ROBBINS: Medicine is a business and there are powerful commercial forces involved. There is a lot of money at stake, and the incentives haven't always been aligned with patient well-being.

DR. DEAN ORNISH: I think that is beginning to change with the Affordable Care Act. Now instead of reimbursement by procedure we're seeing reimbursement by diagnosis. When you say, "Here is X amount of dollars to take care of someone who has got heart disease," then suddenly the doctor might be advising doing fewer procedures and helping people to change their lifestyles.

We all know that lifestyle is important in preventing disease. But now we're seeing lifestyle as a treatment. It can often work as well or even better, at a fraction of the cost, with only good side effects.

There is a convergence of forces that makes this the right idea at the right time. The limitations of high-tech medicine are becoming clear. Stents and angioplasties don't work for stable patients, and the surgery for prostate cancer isn't really necessary most of the time. Meanwhile, the power of these very simple, low-tech, low-cost interventions like lifestyle changes have become increasingly well-documented.

The opportunities are really ripe now for industries to realize that a new paradigm of medical care can be much more sustainable, and even profitable.

Health care costs are reaching a tipping point. They are not financially sustainable for the government nor for many families. Most large businesses are self-insured, and this is coming right off their bottom line. There is a debate between some people who say, "Let's just raise taxes and let the deficit go up," and other people who say, "No, let's just dismantle or privatize Medicare."

I say, "If 75% of the $2.8 trillion in health-care costs are for chronic diseases that can often be prevented and even reversed by making comprehensive lifestyle changes, this can be a third alternative. By teaching people how to change their lifestyles, we can make better care available to more people at significantly lower costs—and the only side-effects are good ones."

2

Caldwell Esselstyn, M.D.
You Can Prevent and Cure Heart Disease. Period.

Caldwell Esselstyn, M.D., is author of the bestselling book, Prevent and Reverse Heart Disease. *Drawing on the insights from his decades of rigorous research and more than 150 scientific papers, Dr. Esselstyn explains, with irrefutable, scientific evidence, how we can literally end the heart disease epidemic forever by changing what we eat. His work is featured in the extraordinarily popular documentary,* Forks Over Knives.

Do you or anyone you know suffer from heart disease? It's the leading killer in the world. Would you like to know Dr. Esselstyn's simple and life-saving prescription?

JOHN ROBBINS: You conducted a study that rocked the medical world. You have said that patients in your study who were compliant, and 95 percent of them were, became virtually heart-attack proof. That is strong language, and your research seems to support it. Yet the idea for your study was born decades earlier.

DR. CALDWELL ESSELSTYN: My work was kindled by reviewing the global literature on cardiovascular illness. It is quite striking that even today if you are a cardiac surgeon and you are going to set up your practice in rural China, Central Africa, or with the Tarahumara Indians in Northern Mexico, you might as well just forget it. You'll make more money selling pencils, because you are not going to have any cardiovascular disease to treat. There is none. These cultures, by heritage and tradition, are fully plant-based.

Yet by way of contrast, when we looked at the autopsies of our 20-year-old GIs in Korea and Vietnam, fully 80 percent already had gross evidence of coronary disease that you could see without a microscope. That study was repeated about forty years later in 1999. This time it was done in the United States, looking at young women and men between the ages of 17 and 34 who had died of accidents, homicides, and suicides. This time the disease was ubiquitous. Everybody at that young age already had the foundation of coronary disease. So it is very discouraging to think that when you graduate from high school in the United States today, not only do you get a diploma, but you also get a foundation of coronary disease.

This is further accentuated by a very interesting phenomenon that occurred during World War II when the Axis powers of Germany overran the low countries of Holland and Belgium and they occupied Denmark and Norway. It was characteristic that the Germans would take away the livestock from these cultures—specifically their cattle, sheep, goats, pigs, and turkeys. So now suddenly these Western-European nations were deprived of animal food and dairy during the war years. In 1951 it was quite striking to see the report in *The Lancet*, England's premier medical journal, by Doctors Strom and Jansen who reviewed the Norwegian experience with heart attacks and strokes during those war years. It was striking that from 1939 to 1945, deaths from stroke and heart attack in Norway plummeted. And yet, as soon as there was a cessation of hostilities, immediately back came the meat, back came the dairy, back came the heart attacks, and back came the strokes.

JOHN ROBBINS: What propelled you to conduct the study you did?

DR. CALDWELL ESSELSTYN: In the late 1970s and early 1980s, I was chairman of Cleveland Clinic's Breast Cancer Taskforce. I realized no matter how many women I helped with breast surgery, I was doing absolutely nothing for the next unsuspecting victim. That led me to a bit of global research. It was striking to find that breast cancer in Kenya was thirty to forty times less frequent than in the United States. In rural Japan after World War II, breast cancer was very infrequently identified and yet as soon as the Japanese women would migrate to the United States, by the second and third generation, they now had the same rate of breast cancer as their Caucasian counterparts. Perhaps even more striking was cancer of the prostate. In 1958 in the entire nation of Japan, how many autopsy-proven deaths were there from cancer of the prostate? Eighteen! By 1978, twenty years later, they were up to 137, but that still pales in comparison to the 28,000 who will die of prostate cancer this year in the United States.

At about that time I made a decision to focus on the leading killer of women and men in Western civilization, which is coronary heart disease. It was apparent that there were multiple cultures that were plant-based where this disease was virtually nonexistent. I thought how exciting it would be if we could help people to eat in a way that would save their heart. Because if they were eating to save their heart, then they would probably also be saving themselves from the common Western cancers of breast, prostate, colon, and pancreas.

In the summer of 1985, I went to our Department of Cardiology at Cleveland Clinic and asked for about twenty-four patients who were ill with coronary artery disease. Twenty-four patients was the maximum number I could manage and still carry out my surgical obligations.

JOHN ROBBINS: So you were given twenty-four patients, most of whom were not doing too well.

DR. CALDWELL ESSELSTYN: As my late brother-in-law said, they were the walking dead. They had failed their first or second bypass; they had failed their first or second angioplasty. They were too sick for these procedures or they had refused them, and five were told by their expert cardiologist that they would not live out the year. I am happy to say that all five of those went beyond twenty years, and all patients who were compliant were able to arrest their disease and we often would see a marvelous reversal of it. This was extremely exciting and very rewarding, because some of those actually occurred before the invention of statin drugs. So we recognized the power of correcting nutrition.

JOHN ROBBINS: The power of nutrition, as you showed, is phenomenal to correct and sometimes reverse disease. Yet most medical schools hardly teach nutrition at all. The philosophy seems to be "a pill for every ill." There is very, very little effort to support physicians in learning about nutrition and its power. Having been at this as long as you have, how would you say the battle is going overall?

DR. CALDWELL ESSELSTYN: Nothing is going on in cardiovascular medicine today with any of the drugs, with any of the imaging, with any of the stents or the procedures or the bypass operations, to treat the causation of the illness. Even my wonderful friends who are cardiologists and cardiac surgeons will concede that these procedures do nothing to deal with the underlying problem. They are just a stop-gap patch job.

There was a normal, and I think probably an appropriate, reticence with Dr. Ornish and me when we did our earlier studies. The medical community was willing to concede that we had demonstrated proof of concept but they felt very strongly that it would be impossible to get large numbers of people to change. But we are just now summarizing a group of more than 226 patients that we have counseled over the past ten years or so. These patients have come to see us from across the United States, and 198 of them had severe coronary artery disease. We have had about a 90 percent compliance rate. The number of new major cardiac events (death, heart attack and stroke) is 40 times less frequent than other representative studies. We have had in our group of compliant patients a total of just one event. This was one patient who had a mild stroke. That means that more than 99 percent had no new major cardiac events.

Many people wonder how we have achieved that degree of compliance. Most cardiologists don't doubt that there would be benefit from our program—they just don't think patients would stick with it. But we have proven that it can be done.

JOHN ROBBINS: How did you do it? I have heard so many people say, "Well, you just can't ask people to be that restrictive in their diet."

DR. CALDWELL ESSELSTYN: The first thing you have to do is to show patients respect. The way that a physician can show a patient respect immediately is by giving them their time. At the Cleveland Clinic Wellness Institute where I direct the Cardiovascular Disease Prevention and Reversal Program, we conduct a single five-hour intensive counseling seminar for ten or twelve participants at a time. Their spouses or partners can come for free. We explain to them in language that either a CEO of a company or a high school dropout is going to be able to understand exactly what has been the cause of their disease and exactly what it is that they can do to be empowered to halt and reverse this disease. In addition to this, we have a very hefty notebook that we give them with a copy of every one of my PowerPoint slides, another forty-four-page handout with additional concepts, and many more recipes to add to the 150 in our book, which we give them as well. We also incorporate an hour and a quarter presentation from a woman who has had twenty-eight years of experience acquiring and preparing plant-based nutrition, dealing with reading labels, dealing with travel, and dealing with restaurants. Then we give them a copy of a DVD of an entire seminar that we recorded earlier. We usually have a presentation from somebody who is local or regional who had a previous successful experience to share their story so that those who are there can say, "Listen, if he or she can do this, I can do it."

We give them a delightful plant-based lunch and stay in touch through email or phone calls as is necessary. What we really want to do is to make this the most significant interchange that they have ever had with a caregiver. I need the five hours. I need the same amount of time that the cardiac surgeon has, but I want it with the patient awake.

JOHN ROBBINS: Currently, 45 percent of Medicare spending is in cardiology.

DR. CALDWELL ESSELSTYN: Those tax dollars are being spent on drugs, imaging, procedures, and operations that have absolutely nothing to do with the causation of the disease. So a poor patient today with a cardiovascular illness is going to have their first stent, their second stent, maybe their third, fourth, or fifth, and then a bypass, and then more stents. Eventually they'll have congestive heart failure and sadly end up dying of a food-borne illness that was never treated.

Right now what we are doing in medicine is really sort of insane. A few years ago, it was disappointing that they didn't teach nutrition in medical school. Quite frankly, at this point it is disgraceful.

JOHN ROBBINS: Your work was featured in the popular documentary, *Forks Over Knives*. The film basically examines the claims that most, if not all, of the degenerative diseases that afflict us can be controlled or even reversed by shifting away from our present diet of animal-based and processed foods. What do you hope people take away after seeing this film?

DR. CALDWELL ESSELSTYN: I think that the film is powerful and enlightening because it is all evidence-based. It is so compelling when you see the science that whole food, plant-based nutrition trumps everything else. It is the absolute cornerstone of what I feel can be a revolution in health.

Even in their teenage years people have coronary artery disease. Yet the human frame is so remarkable that it will go on for several more decades doing the very best it possibly can to resist this assault. Eventually, the assault on the vasculature gets to a point where the body can no longer protect against the development of these plaques and you see the rupture and clinical disease that have taken decades to develop.

The same thing is true, of course, of what happens with hypertension. You don't suddenly wake up as a teenager eating horribly and have hypertension. No, you keep that up, and eventually your body can no longer keep your blood pressure normal. It starts to be elevated. The same thing is true with diabetes. Interestingly enough, the same thing is true of dementia. Why is it that at age 85, 50 percent of Americans or Swedes have dementia? You don't wake up at 85 suddenly with dementia. You have worked doggone hard all through your life to really mess up the vasculature to the brain as well as all the beta-amyloid and the tangles that occur in Alzheimer's. You have to work very hard to set the ground work for that to happen.

It is so exciting to think that this seismic revolution will not only get the United States out of debt, but it will remarkably enhance the lives of so many.

JOHN ROBBINS: One of the things that strikes me about your work is that you are asking people to change something basic, which is the food that they eat. People don't always like that. What keeps you going in the face of opposition and resistance?

DR. CALDWELL ESSELSTYN: Well I think the tenacity of purpose comes from years ago when I used to do some athletics in college. I was on a boat crew, and we actually won a gold medal in the 1956 Olympics. We had a motto, and that motto was "press on, regardless."

3

Neal Barnard, M.D.
Eating to Thrive

Neal Barnard, M.D., is one of America's leading advocates for health, nutrition, and higher standards in research. He is the president of the Physicians Committee for Responsible Medicine, and an Adjunct Associate Professor of Medicine at the George Washington University School of Medicine in Washington, D.C. The author of dozens of scientific publications, Dr. Barnard has also written fifteen books, including the New York Times *bestseller,* 21-Day Weight Loss Kickstart: Boost Metabolism, Lower Cholesterol, and Dramatically Improve Your Health.

Find out what Dr. Barnard has to say about the latest findings in diabetes treatment, some of which come from studies he has personally directed. Hint: The most powerful tool isn't a drug, it's the food on your plate. Then, get his take on cancer, soy, weight loss, and many other crucial topics of our day.

JOHN ROBBINS: You grew up in a North Dakota family of cattle ranchers and doctors. When you look back now, do you feel that those early experiences helped shape your current view of the world and your work?

DR. NEAL BARNARD: My grandpa was a cattle rancher, and his father was a cattle rancher, and his father was a cattle rancher. And my own father was too, except that he really did not care for the cattle business and soon got out of it. He ended up going to medical school and then spent the later part of his life treating diabetes in Fargo. But we still ate like we were in the cattle business. I have vivid memories of bringing cattle to slaughter and that kind of thing. Now, all of those people were good, decent folks, and I am sympathetic to how people have gotten into these walks of life. At the same time, we can clearly do better. Science has moved on and shown us that there is a better path.

JOHN ROBBINS: Speaking of science moving on, you were awarded a $350,000 research grant from The National Institutes of Health a few years ago to study the effect of a low-fat, vegan diet on diabetes. The study results, which you published in *Diabetes Care*, the peer reviewed journal, found that a low-fat, vegan diet did improve glycemic and lipid control and actually did so with greater results than those achieved on a diet that was based on the American Diabetes Association guidelines. Has the Diabetes Association made any changes in their guidelines as a result of your findings?

DR. NEAL BARNARD: When you look around the world, people who tend to not get diabetes don't follow the sort of typical diabetes diet that American doctors hand to patients. A typical clinically recommended diet says that patients should avoid carbohydrates: "Don't eat bread, don't eat pasta, don't eat sweet potatoes, don't eat rice, etc." But if you look in Japan and China in decades past, rice was a staple, and they were the thinnest, healthiest, longest-lived people on the planet—that is, until burgers and cheese came in. Then as rice consumption fell and fat intake rose, diabetes rates skyrocketed.

What the NIH funded us to do was to test an entirely plant-based diet, loaded with vegetables, fruits, whole grains, and beans. It was very low in fat, and the animal products were eliminated. Several things happened. People lost weight very well, and their cholesterol levels improved. But what the diabetes researchers especially noticed was that their blood sugars improved so much that many of them ended up reducing their medications and in some cases, diabetes was no longer even detectable.

In 2009, the American Diabetes Association began citing our research studies in its clinical practice recommendations, and we are grateful for that. We hear every day from people whose diabetes is improving dramatically. More importantly, when people make dietary changes, they may be able to prevent this disease from starting in the first place.

JOHN ROBBINS: My father, who as you know was the founder and for many years the owner of the Baskin-Robbins (31 Flavors) ice cream company, developed a very serious form of diabetes in his 70s, and the prognosis was very poor. But as a result of reading my books and similar ones, he made some major changes in his diet and his diabetes went into remission. He no longer needed injections of insulin or even diabetic pills. The amputation of a foot or a leg that had been envisioned wasn't necessary, and he lived many more good years. He didn't go as far with his dietary changes as you or I might have ideally wanted, but he made major ones. I thought, if a person at that age with such an investment in the lifestyle that he maintained could make those changes and experience dramatic health results in a positive way, then perhaps there is hope for just about anybody.

DR. NEAL BARNARD: Your father was very lucky to have you looking out for him. And we have seen the same sort of success many times. I vividly recall a man who came into our research study and told me about his family history. He had diabetes all up and down his mother's side and his father's side, and his own father was dead at age 30. This young man was 31 when he got his diabetes diagnosis. He came in to see us five or six years later. We put him on a plant-based diet. He was thrilled to do it, and he told us it was much easier than the diets he had been prescribed before. Because he could eat as much as he wanted, and he felt good and energetic. He lost about sixty pounds over the course of a year, at which point the diabetes was no longer detectable.

Another man in the study had had diabetes for nearly twenty years and was on injections several times a day. He suffered from a complication called diabetic neuropathy, which is pain in the feet and ankles that comes from the nerves being attacked by the disease. I remember him telling me how miserable this had been, and that life was hardly worth living. After about five or six months on a healthy, vegan diet, he came in to say, "You won't believe this. My neuropathy is completely gone!"

I followed with him for years because I had never heard of this happening before. In medical school we were taught that diabetes is a one-way street. But for this man, the neuropathy never came back. And we have seen this for many people since that time. What we discovered is that diet changes really are powerful. They vary from person to person, of course, but it is wonderful to see what can happen when a person puts a healthy diet to work.

JOHN ROBBINS: You founded Physicians Committee for Responsible Medicine (PCRM) in 1985. What does your organization do?

DR. NEAL BARNARD: We promote preventive medicine, especially good nutrition. We also do clinical research studies. We have done quite a lot on the role of nutrition in diabetes, weight loss, and cholesterol management, as well as a number of studies looking at migraines, arthritis, and menstrual pain. We have done studies on the applications of dietary changes, such as on how people can make nutritional changes at the workplace or in doctors' offices. We also want research to be done better, meaning more ethically. So we promote alternatives to the use of animals in research.

JOHN ROBBINS: When you publish your research in medical journals, have you ever been challenged for your advocacy of a vegan diet on the grounds that you may be concerned for the animals, regardless of the actual health data?

DR. NEAL BARNARD: No, and there are a couple of reasons for that. First of all, let me say that concern for animals is a really good thing. I wish everyone were motivated by compassion, because that would make for a better world. And for myself, I wish that I had a keener, ethical sense earlier in life, because I have often reflected about the times when I drove cattle to slaughter or went hunting as a kid. We did a lot of things that, now that I am a little older and maybe wiser, I really wouldn't do again.

Having said that, when we do a clinical trial, it goes through a rigorous process of independent peer review. Our statistics are all done by people who are unbiased and are masked to the specifics of the research. When they are evaluating blood tests and other results, they don't know who is in which diet group. So when the NIH or other funders or reviewers look at our data, they can see that they are without bias. As a result, our diabetes trial yielded six or seven separate research publications. The American Diabetes Association published our first results in its journal, followed by *The American Journal of Clinical Nutrition, The Journal of the American Dietetic Association*, and several others.

With that said, sometimes a broader social perspective can be helpful. People come in to see us because they want their diabetes to get better or they want to lose weight or to get their cholesterol down. But after they have been on a plant-based diet a little while, many end up saying, "You know, I read an article by somebody named John Robbins." Or they read other enlightened authors, and they say, "The diet that is good for the environment is not much different from the diet that is good for my coronary arteries." Or they read something about animals. They have these "ah-ha!" moments that I am glad to see, because it means they have that much more motivation for staying on a healthy regimen.

JOHN ROBBINS: I actually find it a source of gratitude that the same food choices that are healthiest for us; that give us the strongest immune systems and the healthiest, longest lives; and that lead to the least susceptibility to diabetes, heart disease, and cancer, are also the kindest to other animals and the best for the environment. The fact that there is that coherence doesn't jeopardize or cast any kind of aspersion on the diet, it actually strengthens it and it shows that we are connected to the world in which we live in ways that we sometimes aren't even aware of.

DR. NEAL BARNARD: Yes. And think back a few years. In the late 1950s and early 1960s it was becoming quite clear that tobacco caused lung cancer. The researchers doing that work found themselves absolutely convinced. And yet they still studied smoking in order to try to tease out what is it in the cigarette smoke that causes cancer, what happens to people over time as they develop lung problems, or other aspects of the tobacco-health puzzle. In the process, most cancer researchers stopped smoking. The same is happening now, a generation later, with diet.

Now researchers are saying, "Wait a minute. Vegetables really are good for you, and so are fruits. And so is getting away from cholesterol, and, for that matter, from animal products in general." There comes a time when research findings are so overwhelming. We have seen that, for example, with hot dogs and other processed meats and colorectal cancer. It is beyond dispute that these foods contribute to colorectal cancer. You simply can't be open-minded about it anymore because now the evidence is just open and shut. Even so, there are still open questions, and we still have to do our research in as unbiased a way as possible, and we have ways to eliminate bias among those who analyze the results.

JOHN ROBBINS: Well speaking of research and public confusion, you and I have both been highly critical of The Atkins Diet. When Dr. Robert Atkins died, PCRM obtained and released a medical report on his death and there was an ensuing controversy. What actually happened there and most important, how does this reflect on the health consequences of the diet that Dr. Atkins followed and advocated?

DR. NEAL BARNARD: Well it was quite a time. The Atkins books were at the top of the bestseller lists, and Dr. Atkins encouraged people to believe that they could safely eat high-cholesterol foods and high-fat foods without risk. All the while, he was hiding from the public that he had cardiovascular disease himself. When he died, his autopsy report laid out the facts. A cardiologist who lived in the Midwest asked for a copy from the coroner, because he felt it was important for people to know the truth. So he sent it to us and we released it to the press. People did, with some justification, ask why we were talking about a dead man's medical condition. I made the judgment that it was critical for those tens of millions of people who imagined they could safely eat gravy and bacon to know that they really were putting themselves at risk. Unfortunately much of the residue of that diet has continued. People still, in many cases, think there is something wrong with bread or rice and that there is something safe about pork rinds. That is a terribly dangerous lesson for people to have memorized.

JOHN ROBBINS: There are many people who are afraid that soy products are unhealthy. Can you share some of your insights?

DR. NEAL BARNARD: The Internet is a wonderful thing. It shares information very quickly, but it shares misinformation just as quickly. Many decades ago, researchers became aware that soy products and many other foods in the legume (or bean) category contain substances called isoflavones. If you were to draw the chemical structure of a typical isoflavone on a blackboard, it would look somewhat similar to estrogen—female sex hormones. Isoflavones are not estrogens, but they look somewhat similar. So some people then worried that, if isoflavones really did act like female sex hormones, then men consuming soy products might become effeminate or have a lower sperm count, or women consuming soy might be at risk for cancer. Researchers have studied these things, and it is very clear that in parts of the world where men consume large amounts of soy—for example in China or Japan, where tofu and soy milk have long been consumed—there has been no problem with fertility.

Studies following girls who consume soy products during adolescence, when the breast tissue is forming, have shown that their risk of developing breast cancer is actually about 30 percent less than that for women who did not consume soy. So, if anything, soy products are helpful.

We now have three research studies looking at what happens when women consume soy after they have had breast cancer—two in the United States, and one from China. They show that women who consume tofu, soy milk, or other soy products have about a 30 percent reduction in their risk of the cancer coming back. So in other words, soy seems to be protective. Having said that, soy products are totally optional. A healthful plant-based diet could follow a Mediterranean pattern based on vegetables and fruit and pasta. It could follow a Latin American pattern with beans and tortillas. Or it might follow an Asian pattern, which is where soy products come in. Soy does not increase cancer risk. It does quite the reverse, but you don't have to have it if it is not your thing.

JOHN ROBBINS: Cancer is a terrifying prospect in a lot of people's lives, and rates have been increasing. You founded The Cancer Project. What is the organization doing, and why is its work so important?

DR. NEAL BARNARD: This is near and dear to my heart. My father was diagnosed with prostate cancer when he was in midlife and he had a miserable time with it. So many people know nothing at all about what leads to cancer or what might help prevent it. But we have abundant evidence that foods play a big role. The Cancer Project offers nutrition and cooking classes in cities across the United States. If you live in Pittsburgh or Los Angeles or Seattle or in one of a hundred other cities, you can come on in and there will be others just like you who want to know about how foods can protect you against cancer or help after diagnosis. You will learn about nutrition and cancer, and will have a chance to stir the spaghetti sauce, and can even bring your reluctant spouse in too, so that we are all learning together. We have instructors, we have books, we have videos, and we have a website, which is *www.cancerproject.org.*

My aspiration is that instead of only hoping for research to find a cure, or just hoping that a mammogram or a PSA test will find your cancer early enough, we can go another step. I hope we can use foods to prevent cancer so that it never has to happen in the first place.

JOHN ROBBINS: You have been at this for many decades now. When you look back on these years and on your work, what has surprised you?

DR. NEAL BARNARD: Everything surprises me. It surprises me that Bill Clinton, a President who was known for jogging to the nearest fast-food chain and eating junk food, and who was gaining weight and looking less and less healthy, finally decided to change. He not only adopted a plant-based diet, but he decided to tell the whole world. He went on television saying that the change was not so hard, and encouraging others to give it a try.

It surprises me that in almost every school in America, students are asking for vegetarian or vegan choices.

But it also surprises me that the people working for the meat, dairy, and junk-food industries persist not only in producing unhealthful products, but also in fighting efforts to change what is in school lunches, and what is in the food stamp (SNAP) program. Even though their own families are at risk and they themselves are paying a price for unhealthy eating habits, they fight to keep the worst foods front and center on the American plate.

So frankly, everything surprises me. But what can you do? Life is short. You have to just work as hard as you can to get the word out. And over time people really do take this knowledge in hand, and they share it with other people they know. The most important thing is that they share it with their kids, because that will change the fundamental direction in which we're headed.

JOHN ROBBINS: Many things that I had once thought were controversial have over time become mainstream. Things that had been bitterly fought are now being taken as self-evident. At this point in history, at this point in the arc of change, what steps would you like to see people take?

DR. NEAL BARNARD: We need to think societal change. The Internet enables us to reach many, many people with the click of a button. That is great, and we need to use it. We also need to leverage the power of business. We have been working with a number of businesses now, including the GEICO insurance company that is known for its cute, green lizard icon. GEICO instituted plant-based diets at ten of its thirteen facilities around the United States, so that their employees could try it, taste it, and adapt to healthier meals. And we have tracked the results. In our first GEICO study over the course of 22 weeks, employees lost an average of 11 pounds, and saw their blood-pressure levels drop. They also missed less days from work. We published those findings in the *American Journal of Health Promotion*.

We need to work for societal change, because the price of bad food habits is simply too high. Americans now eat more than a million animals every hour. And many more are consumed elsewhere around the world. The environment is degrading faster than many of us had ever thought, and human health is paying a terrible price for our current shortsightedness.

JOHN ROBBINS: What is the food revolution that you would like to see take place?

DR. NEAL BARNARD: My hope is that what is on the plate will be different and that what is growing in the fields will be different, as well. In turn, animals will have an entirely different experience—that is, they will no longer be considered dinner. When I went downtown yesterday to a meeting, I went into a large office building. There was a man outside finishing up his cigarette before he could go into his nonsmoking building. And I hope that, ten or twenty years from now, that same guy will be standing outside finishing up his chicken wing before he is allowed to go into his vegan office building.

JOHN ROBBINS: There are so many people today waking up to the reality that the industrial food machine is spewing out and advertising to our kids some of the most unhealthy foodlike substances imaginable. The price we are paying for it as individuals, as families, and as a society, is exorbitant. Yet so many of the policies of our government have reinforced and supported the industrial food machine: feedlot agriculture, Monsanto, McDonald's, factory farms, and even now, genetically engineered foods. In the face of the momentum of that and all the lobbyists down in Washington, what are the steps that a group of people can take to make a difference politically?

DR. NEAL BARNARD: Every five years, Congress decides what the agricultural subsidy programs are going to look like as it formulates the Farm Bill. Up until this point, the U.S. Government gives almost nothing to vegetables and fruits, and it gives huge subsidies to animal agriculture. The biggest part of the Farm Bill is the SNAP Program— the Supplemental Nutrition Assistance Program, which used to be called Food Stamps. The program is well intentioned—it is supposed to provide food for needy people. However, the program has become largely a service for the junk-food industry. Other food-assistance programs provide reasonably healthy food. The WIC program, which serves women, infants, and children, is limited to a finite list of foods that are more or less healthful. School lunches have at least some limitations now as well. For example, you can't serve sugary sodas in most school lunches. The SNAP program is different and is not remotely health-conscious. It does cover vegetables and fruits and grains and beans, but it also covers sausage, cheese, Red Bull, candy, potato chips, and sugary sodas. The fact is, convenience-store operators who operate in neighborhoods in "food deserts" are reimbursed just as well for junk food as they are for fresh fruits and vegetables. As a result, they have no incentive for stocking anything other than candy and packaged snacks. They are not going to put an orange or some fresh spinach on the shelf because fresh things have a shorter shelf life. If the food stamps pay for potato chips and other things that have a longer shelf life, economic pressures favor the worst foods. Small wonder that economically disadvantaged people are at higher risk for obesity and diabetes, compared with their wealthier counterparts.

We would like to limit the SNAP program to those foods that are healthy: grains, beans, vegetables, and fruits, whether they are fresh, frozen, or perhaps in a can. If the retailers were only compensated with government money for healthful foods, it would spell the end of food deserts. The result would be that needy folks who are now paying a terrible price for the junk food avalanche that is all around them could instead become the healthiest people in America.

4

Dr. T. Colin Campbell
Is Animal Protein Good for You?

T. Colin Campbell, Ph.D., has been at the forefront of nutrition research for more than forty years. His legacy, The China Project, is the most comprehensive study of health and nutrition ever conducted. Dr. Campbell's academic credentials are extraordinary. He is the Jacob Gould Schurman Professor Emeritus of Nutritional Biochemistry at Cornell University. Dr. Campbell has more than seventy grant-years of peer-reviewed research funding and more than 300 research papers on his resume. He is coauthor of the bestselling book, The China Study: Startling Implications for Diet, Weight Loss and Long-term Health.

Dr. Campbell grew up on a dairy farm. Over the decades, his research has led him to believe that dairy products, and animal protein in general, are having a profound impact on human health that is not at all what most of us imagine. His decades of research have brought him to some startling conclusions.

JOHN ROBBINS: Dr. Campbell, like you I grew up eating a lot of dairy products. How has your research impacted your personal dietary choices?

DR. T. COLIN CAMPBELL: I was raised on a dairy farm and milked cows until starting my doctoral research more than fifty years ago at Cornell University in the animal-science department. Meat and dairy foods were my daily fare, and I loved them.

When I began my experimental research program on the effects of nutrition on cancer and other diseases, I assumed it was healthy to eat plenty of meat, milk, and eggs. But eventually, our evidence raised questions about some of my most-cherished beliefs and practices.

Our findings, published in top peer-reviewed journals, pointed away from meat and milk as the building blocks of a healthy diet, and toward whole, plant-based foods with little or no added oil, sugar, or salt.

My dietary practices changed based on these findings, and so did those of my family.

JOHN ROBBINS: What did you discover?

DR. T. COLIN CAMPBELL: In human population studies, rates of heart disease and certain cancers strongly associate with animal-protein-based diets, usually reported as total fat consumption. Animal-based protein isn't the only cause of these diseases, but it is a marker of the simultaneous effects of multiple nutrients found in diets that are high in meat and dairy products and low in plant-based foods.

Historically, the primary health value of meat and dairy was touted to be a generous supply of protein. But therein lay a Trojan horse.

More than seventy years ago, for example, casein (the main protein of cow's milk) was shown in experimental animal studies to substantially increase cholesterol and early heart disease. Later human studies concurred. Casein, the properties of which, it's important to note, are associated with other animal proteins in general, also was shown during the 1940s and 1950s to enhance cancer growth in experimental animal studies.

Casein, in fact, is the most "relevant" chemical carcinogen ever identified; its cancer-producing effects occur in animals at consumption levels close to normal—strikingly unlike cancer-causing environmental chemicals that are fed to lab animals at a few hundred or even a few thousand times their normal levels of consumption. In my lab, from the 1960s to the 1990s, we conducted a series of studies and published dozens of peer-reviewed papers demonstrating casein's remarkable ability to promote cancer growth in test animals when consumed in excess of protein needs, which is about 10 percent of total calories, as recommended by the National Research Council of the National Academy of Sciences more than seventy years ago.

JOHN ROBBINS: Do you see any beneficial role for bovine dairy products in the human diet?

DR. T. COLIN CAMPBELL: I see no redeeming value in consuming dairy products from a nutritional perspective. The dairy industry has long promoted the myth that milk and milk products promote increased bone health—but the opposite is true. The evidence is now abundantly convincing that higher consumption of dairy is associated with higher rates of bone fracture and osteoporosis, according to Yale and Harvard University research groups.

Plant-based foods turn out to have plenty of calcium along with far greater amounts of countless other essential nutrients (such as antioxidants and complex carbohydrates) than meat and dairy.

The dairy industry has promoted its products as a good source of high-quality protein. But higher-protein diets achieved by consuming animal-based foods increase the risks of cancer, cardiovascular diseases, and many similar ailments.

Protein consumed in excess of the amount that we need, and most of us do consume more than we need, actually has some pretty serious consequences.

I wouldn't have expected to see this because, like you John, I was raised in a dairy family. I always believed that the good old American diet was about as good as it gets. And subsequently I went off to graduate school and actually did my doctorate dissertation expecting the research to prove my beliefs to be correct. I never imagined that we would find the things we did find. It turns out that protein from cow's milk is also a pretty potent inducer of higher cholesterol levels, atherogenesis which is the forerunner for cardiovascular disease, and a number of other sorts of illnesses.

JOHN ROBBINS: It would seem that cow's milk is nature's most perfect food for a baby calf—who has four stomachs and is a ruminant animal that will gain about 200 pounds in its first year. I guess if a human infant has those characteristics, it would probably be the right food for that child. I think the cultural belief system that holds dairy products as exemplary is causing a lot of damage to people. The mainstream belief is that the saturated fat in dairy products and other animal foods can contribute to heart disease. But, your studies and many others indicate that many of the chronic diseases found today result from the consumption of animal protein.

DR. T. COLIN CAMPBELL: Some of the most compelling evidence of the effects of meat and dairy foods arises when we stop eating them. Increasing numbers of individuals resolve their pain (arthritic, migraine, cardiac) when they avoid dairy food. And switching to a whole-food, plant-based diet with little or no added salt, sugar, and fat, produces astounding health benefits. This dietary lifestyle can prevent and even reverse 70 to 80 percent of existing, symptomatic disease, with an equivalent savings in health-care costs for those who comply.

This treatment effect is broad in scope, exceptionally rapid in response (days to weeks), and often, lifesaving. It cannot be duplicated by animal-based foods, processed foods, or drug therapies.

By contrast, any evidence that low-fat or fat-free-dairy foods reduce blood pressure is trivial compared with the lower blood pressure obtained and sustained by a whole-foods, plant-based diet.

Based on the scientific evidence, and on the way I feel, I know beyond any doubt that I am better off for having changed my diet to whole and plant-based foods.

JOHN ROBBINS: We have compelling evidence that many of our chronic and devastating illnesses can be prevented with improved nutrition. But for someone who actually has cancer, do you see diet as having a role in effective anti-cancer treatment?

DR. T. COLIN CAMPBELL: Quite possibly. For a long time, we talked about nutrition as it might relate to the prevention of future problems. But now in recent times we have been see-ing that the same diet that tends to prevent future problems can also be used to reverse and treat certain illnesses after they are present. With heart disease and Type 2 Diabetes, the role of diet in reversing disease once it is present is very clear. With cancer, the preventive role of nutrition is solidly documented. But for reversal of existent disease in the case of cancer, the evidence is not as strong, although it does exist.

We need more good-quality research on the question concerning the effect of diet, and protein in particular, on the development of cancer. But unfortunately in our medi-cal community, that kind of research has not really been done very well. The reason is that doctors have generally not been schooled in nutrition and they are extraordinarily reluctant to admit that this is a good idea.

JOHN ROBBINS: If you could design the diet that our governments would advocate and support, based on what is healthiest for people, what would the diet look like?

DR. T. COLIN CAMPBELL: I like to suggest a "whole foods, plant-based diet." By that I mean whole vegetables with lots of colored vegetables included. Whole fruits, whole grains, and legumes, as close as possible to the way they have been prepared by nature, without adding a lot of oil or fat or sugar. I also believe it is optimal to consume much of our food in its raw form. By this I am talking about salads, fresh fruits, and fresh vegetables. I'm not a proponent of a 100 percent raw-food diet for a variety of reasons, but I believe that an abundance of raw foods is a really smart way to go.

JOHN ROBBINS: What is the relationship between animal food and the alkalinity or acidity issues in the body?

DR. T. COLIN CAMPBELL: Well, animal food in general creates a sort of metabolic acidosis that can drop the body's pH one or two tenths of a unit. That can make quite a difference. The animal foods can create a more acidlike environment, which has been shown to have some significant effects on enzyme activities.

JOHN ROBBINS: It seems to me that a great number of people are suffering from some subclinical levels of acidosis. If someone is wanting to alkalize their system, how would you suggest they go about that? What foods should they eat? Are there certain activities they should undertake or avoid?

DR. T. COLIN CAMPBELL: I don't think we really have good, solid research on that question. But what we do know is that plant foods will create a more alkaline environment in our bodies and the big contribution to acidification is simply animal-based foods.

I was asked recently for my view on the use of alkaline water and some of these products that are supposed to add alkalinity in the human body. I really am sorry, but I can't give a good answer to that. There may be some research, I just haven't been familiar with anything that convinces me yet. I am open to the possibility of alkalizing foods or maybe alkalizing water, but I don't think we have enough data to know for sure.

JOHN ROBBINS: It seems that you are saying that the central thing to do is to derive as much of our nutrients, our proteins, our vitamins, our phytochemicals, our fats, and our carbohydrates from plant sources as possible.

DR. T. COLIN CAMPBELL: Absolutely. I think a lot of the time in the field of medicine and nutrition research in particular, we get caught up in the details rather than thinking about the big story. I think the really big story is just getting some wonderful vegetables and fruits and grains in their natural and whole state, and learning how to use them and adapt to the taste of them.

JOHN ROBBINS: You have been a researcher, a scholar, and a scientist for more than forty years. You went to China and conducted the most comprehensive study of human nutrition in world history. Now you and your son have written about what you learned in *The China Study*, which is a runaway bestseller. You have come to be on the front lines, challenging the prevailing nutritional dogma. How is that for you?

DR. T. COLIN CAMPBELL: It has been gratifying and honestly somewhat surprising to see the story of *The China*

Study go as far as it has. I was always quite confident of what we had written and what I had done. But I did not know how the public would respond. Some people are persuaded by the time they get to chapter three, where I am talking about protein and cancer. Other people are persuaded by the last third of the book, where I talk about my twenty years' worth of experience on National Policy Boards on food and health and some of the insights I gleaned along the way.

JOHN ROBBINS: You are running up against government, medicine, corporations, and the media, all of which have at times concentrated on profits at the expense of health. Together they have created a level of confusion about nutrition and have at times stifled and even attempted to destroy viewpoints that challenge the status quo. You, yourself, were almost expelled from a committee of scientists because you dared to suggest that there might be a link between diet and cancer. Where do you find the strength to continue working so hard despite all the obstacles, discouragement, and resistance that you encounter?

DR. T. COLIN CAMPBELL: Well it took many years for me to get enough confidence. Honestly, your work, as well as the work of Dr. John McDougall and others, helped to give me courage and inspiration. All of a sudden it occurred to me that I wasn't standing alone. I was just contributing my part. I get a lot of strength from being part of this community.

We have to go in this direction, because now we have enough evidence from the laboratory, from the field, and even from certain philosophical and environmental resource perspectives. We have enough evidence now to demonstrate it. I have taken my hits and I am sure you have too, some of them pretty serious. But I keep trying to look past them, because this is a story worth telling.

5

Joel Fuhrman, M.D.
Specific Steps to Excellent Health

Joel Fuhrman, M.D., is a family physician and researcher who specializes in reversing disease through nutrition. Millions of people were inspired by his work through the PBS special, 3 Steps to Incredible Health, *which was the network's highest grossing pledge drive program of 2011. Dr. Fuhrman is the number one* New York Times *bestselling author of* Eat to Live, *and research director for the Nutritional Research Foundation. He coined the term* "nutritarian," *which means someone who strives for nutritional adequacy for improved health and whose food choices are high in nutrients per calorie.*

What is a diet that won't just be a little better than the norm, but that will maximize your potential for optimal health? Which vitamins are healthy for you, and which are so hazardous that they should come with warning labels? What about food allergies, and whole grains? The doc is in, and he's here to give you the latest breakthroughs in nutrition for optimal health.

JOHN ROBBINS: I find you somewhat unique among medical doctors. Very few even study nutrition, much less help their patients apply it. It has occurred to me that a doctor who doesn't know about nutrition is something like a fireman who doesn't know about water.

DR. JOEL FUHRMAN: The medical profession developed with a primary focus on developing and prescribing medications to reduce people's symptoms, rather than on dealing with the causes of disease. Thousands of years ago, a doctor was someone who taught people how to live a healthy life. But I think it has evolved to now being a person who is an expert in giving medications. The trouble is that taking toxic remedies to resolve bad lifestyle choices is largely ineffective and allows for peoples' underlying disease process to continue advancing. I think there are a lot of doctors re-evaluating their careers right now. Fortunately I had the opportunity to learn about nutrition at a young age and to pursue a career where nutrition became the centerpiece of my medical practice. It's afforded so much personal reward to help many thousands of people reverse their conditions and get well, without medications.

JOHN ROBBINS: What are the rewards that our readers could look forward to if they were to heed your suggestions?

DR. JOEL FUHRMAN: Proper nutrition is the foundation for protecting yourself from cancer, obesity, diabetes, heart attacks, and strokes. Most people who learn about my Nutritarian approach are in poor health, after living thirty to fifty years on a diet that breeds disease. Now they can lose weight and they can get in better health, lower their blood pressure, or get rid of their diabetes. But the question is: Is that enough? After forty to fifty years of eating a cancer-causing diet, will a change now be sufficient to prevent you from having your life cut short with a tragic cancer at a later stage from what you ate in the first half of your life?

My answer is that eating decently or making moderate beneficial changes is not adequate enough to repair the broken DNA cross-link—the methylation of DNA. In other words, whatever damage occurred to your cells over those years even before you were born, even when your eggs were in your mother's body before you were conceived, has an impact still. Your health can be affected long-term by your exposure to toxins, and by a lack of nutrients.

So we maximize the body's ability to repair dysfunctional DNA, to remove toxins from the cells, and to restore itself and its immune system. We have to undo the damage that, if left unchecked, would lead to cancer. My niche in the nutritional world is to help people who don't just want a little better health, but who want to know what would be optimal. To maximally repair cellular damage, we are going to push the envelope of human longevity and really see if we can win the war on cancer.

My approach is not just about losing weight and not having heart attacks. It's also about maximizing healthy life expectancy. A hundred years ago people lived about as long as they are living now. Life expectancy was lower because we had much higher infant and child mortality rates, and many women died during childbirth. But in the real comparison of health issues, we found that people who lived fifty or a hundred years ago actually didn't have as much disability, discomfort, and pain in the last ten years of their lives as people do today. They mostly had a bad time in the last three to six months of their lives. Now people are over-medicated and sickly, and have a very poor healthy life expectancy. This is because the American diet has degenerated with so much fast food, processed food, and refined food.

We are literally seeing an epidemic of disease that's weighing down our health-care system, weighing down our economy, and creating huge amounts of personal human tragedy. Excellent nutrition is tremendously powerful in giving us personal choice and control of our health destiny. It can also be tasty, fun, and exciting to actually be in great health and continue to have a healthy life as we age.

JOHN ROBBINS: You coined the term "nutritarian." Would it be correct to say that refers to someone whose food choices provide an optimal amount of nutrients per calorie?

DR. JOEL FUHRMAN: Sure, but also a person who strives for better nutrition. A "nutritarian" is someone who chooses to have excellent nutrition, in order to have better health. A lot of people are "nutritarians" and have never heard of the word before. Micro-nutrients are the noncaloric portion of food—the vitamins, minerals, phytochemicals, and other newly discovered factors that are so important for human health. I use the formula $H=n/c$ (health equals nutrients divided by calories). If you want to live a long time, repair cellular damage, and protect against late-life diseases like dementia, you want a large amount of micro-nutrients per caloric buck. That means a high n/c ratio. You need a broad spectrum of micro-nutrient diversity. You need to eat a lot of green vegetables and other colorful foods that supply you with high levels of the phytochemicals that prevent cancer.

JOHN ROBBINS: Which micronutrients are the most important to consume to make sure we have superior immune function?

DR. JOEL FUHRMAN: The most important nutrient for you to consume is the one that you're missing. You have to make sure that you have comprehensive micronutrient adequacy. If your body is an orchestra, you can't have a few extra drum sets pounding on while there are no flutes present. You have to make sure everything's present to create the music.

It's not just about eating more broccoli. It's also about, for example, eating some mushrooms, because mushrooms supply some particular nutrients that are not present in many other foods and that are important to achieve comprehensive micronutrient adequacy. Micronutrient adequacy is not about numbers, but about the complexity and the diversity that is necessary for superior immune function.

JOHN ROBBINS: What are the micronutrients that most Americans are deficient in, and what are good sources to provide them?

DR. JOEL FUHRMAN: The American diet couldn't be better designed to create cancer and heart attacks had we designed it for that purpose. Right now it's degenerated to the point that 62 percent of calories are from refined foods and about 26 percent come from animal products. Of the 10 percent of calories that remains from unrefined plant food, half of that comes from white potato products, which are not exactly nutrient-rich. Americans are just not consuming fruits, vegetables, beans, nuts, and seeds. They're not consuming sufficient quantity of natural plant foods with a broad assortment of protective micronutrients.

I coined another acronym called G-BOMBS, to help people remember the foods that they are supposed to be consuming on a regular daily basis.

The "G" stands for greens, which includes both raw and cooked green vegetables. Most of all we're focusing on leafy greens and cruciferous vegetables like broccoli, cabbage, and Brussels sprouts. A mixture of raw vegetables and cooked green vegetables in your diet has been documented to have an association with longer life and protection against cancer.

The first "B" stands for beans, which are associated with longer life and are a rich source of phytonutrients that link to longevity in humans.

The "O" stands for onions. I've been shocked and pleasantly surprised at the studies that show how protective onions are against cancer. I have changed my diet dramatically in the past five to ten years, trying to shred raw onion on my salad, and to eat more onions and mushrooms.

The third one is "M" for mushrooms, which is a very exciting part of research. It appears that mushrooms have the ability to actually enhance human immune function. They help to label cells that are becoming abnormal for your immune recognition to remove them. Mushrooms also have weight-loss benefits outside of being low in calories, because they actually have angiogenesis inhibition effects, which prevent the blood vessels from fueling fat expansion. So they have fat-inhibiting effects as well as cancer- and tumor-inhibiting effects.

The second "B" stands for Berries. Berries have Polyphenols and Anthocyanidins, which are very beneficial for the brain and the body's immune system.

The "S" stands for Seeds, like flax seeds, Chia seeds, hemp seeds, pumpkin seeds, and sesame seeds. Having some seeds on a regular basis can be beneficial for numerous reasons. The lignins have cancer-protective effects, and the fatty acids are beneficial for health, the stabilization of inflammation, and heart regularity.

We can construct dietary recommendations to maximize human immune function, and to repair and resist the process of aging. We have an opportunity to live longer and in better health than in any other time or place in human history.

JOHN ROBBINS: Is there a role for whole grains in your diet?

DR. JOEL FUHRMAN: Well in the diet style I'm recommending, an intact grain is more favorable than even a whole grain that has been ground into a flour. Certainly there's a role for these things in moderate amounts, but it should be as close as possible to its natural, unprocessed state. So for example black or wild rice would be a better food than rice flour. When rice is processed or ground, the hypoglycemic effect is increased. When you eat an intact rice it's a whole grain, and you're not grinding it all. Same thing with wheat berries over whole wheat, or steel cut oats over oat flour. Even though it might be the same food, it's still more favorable to consume it in the least-processed form possible.

We rate carbohydrates based on certain criteria and that criteria helps us to discern what proportion of those foods would be best in the diet.

Because of their fiber content, their micronutrient content, their resistant starch content, and their lower glycemic effect, the most favorable carbohydrate might be beans. Rice wouldn't be as favorable as beans, but some rice could be included in your diet. Beans, squashes, peas, and lentils would all be more favorable foods, for example, than even brown rice. Certainly I would consider white rice and white potato to be less favorable because of factors like resistant starch, low fiber and nutrients, and high-glycemic effects. We want people to make sure they have enough room in their stomach to consume the full spectrum of nutrient-rich foods.

JOHN ROBBINS: Are some beans preferable to others? What about soy, which of course has a higher fat content and a different nutritional profile, for the most part, than other beans? What do you think about soy?

DR. JOEL FUHRMAN: I think soy is an excellent healthy food. But I also think that nutritional variety is important. I think that it's helpful if we have a few different types of mushrooms in our diet and it's helpful if we have a few different types of beans in our diet, as well. Our diet shouldn't be revolved around the consumption of one bean. I want you to have some lentils, peas, black beans or navy beans, azuki beans or soy beans. They all have different nutritional profiles and different advantages. It's best to have maybe a few different types in your diet on a weekly basis and not just eat a diet that's soy-bean based.

Like any other bean, soy bean is a very wholesome natural food. If it's processed and made into an isolated soy protein-based product, like a soy hot dog or soy bacon, it's not a natural food anymore. Just like corn, which is a pretty wholesome food, but when you make it into corn chips or Fritos it's not going to be a healthy food any more. We have to eat food that is as minimally processed as possible.

JOHN ROBBINS: I wonder why there is so much disagreement among health advocates about the best way to eat. Some of this seems so obvious.

DR. JOEL FUHRMAN: I think it is pretty obvious, but we like to believe the things we were raised on. We formulate opinions very young in life. It can be hard to change people's minds sometimes because they don't make decisions based on science and logic. They make their decisions, especially about food, based on emotions and attachments, and these are hard for many people to break.

I'm suggesting that it's not good enough for people to eat less calories to lose weight. You can't just try to eat less because then you have unrelenting food cravings and food addictions that are counteracting that desire to lose weight and eat less. Your weight just yo-yos and you gain it back again. I'm suggesting that paying attention to the *quality* of what you eat will help mitigate hunger, lessen food cravings, and remove food addiction. Only through enhancing the micronutrient quality of your diet can you lose cravings and physical drives that get intertwined and intermingled with your emotional drive in a primitive brain that wants to continue the addiction.

Diets of all descriptions are doomed to fail when people aren't paying attention to meeting the micronutrient needs of the human body. When we eat a diet with better nutritional quality, all types of benefits that make you healthier will enable you to lose weight effortlessly. You won't crave overeating as much, and eventually you'll be put back in touch with what I call true hunger—where your body becomes a precise computer, telling you the right amount of food you require to maintain your perfect weight. It doesn't need to be a guessing game in which you weigh the food and measure how many calories you need. You can get back in touch with your instinctual drive that tells you how to eat, and you'll enjoy food more. You'll enjoy the taste of it more and you'll be able to never be on a diet. You just eat what you want, except what you want is less and what you desire will be foods that are good for you.

JOHN ROBBINS: What about food allergies? How can people know if they are allergic to certain foods, and what can they do about it if they are?

DR. JOEL FUHRMAN: Some of these problems developed when people were very young—maybe even before they were born. You may have been put at high risk for allergies from folic acid or the diet your mother was eating or what you were eating when you were young in life.

The first step that I want to take with food allergies is of course identifying and avoiding what you are allergic to. The second thing is to achieve excellent nutrition and an excellent balance of everything that might be missing. Supplying sunlight and the right fatty acids and micronutrients helps the immune system function better so the allergies have the potential to get better. It usually takes a few years of actually achieving excellent health to see the allergic tendencies start to diminish. I have much experience with patients who had food allergies, even environmental allergies—including allergies to cats and hay fever—that went away or gradually resolved. Once they are in great health with no nutritional deficiencies, which often takes a few years, then we can try to do oral food challenges and work to see if they can slowly get rid of their allergies. In many cases they just see their allergies start to improve over time with this program.

JOHN ROBBINS: You mentioned sunlight. What do you think about Vitamin D supplements?

DR. JOEL FUHRMAN: I think that Vitamin D supplements are essential for a huge segment of the population, especially those of us in northern climates who are living and working indoors and can't get adequate Vitamin D from the sun. I could tell you a lot of stories. For example, one person came into my office unable to eat or swallow. He vomited up everything he ate. He went to various specialists and one person wanted to do surgery to stretch his pyloric valve. Another person wanted to inject him internally with Botox to relax the pyloric valve because he kept vomiting. I found his Vitamin D to be deficient, which had been going on for years. With fixing his Vitamin D, he recovered in a short period of time. We have an epidemic of Vitamin D deficiency. We need people to have adequate levels of all nutrients in their blood, including Vitamin D, in order to maximize immune function and good health.

JOHN ROBBINS: Folate has been found to be nutritionally beneficial. As a result, many people are taking folic acid regularly in multivitamin supplements, and women are strongly encouraged to take it during pregnancy. What's your perspective on folic acid?

DR. JOEL FUHRMAN: Folate is found in green vegetables and beans. But folic acid is not the same biological compound as folate. I'm suggesting that the data indicates that folic acid might be a powerful contributor to the cancer epidemic. Even when you take it during pregnancy to prevent birth defects it has negative effects on your unborn child. You can't substitute folic acid for the folate that nature intended us to get.

When women who take folic acid are followed thirty to forty years later, they have high rates of breast cancer, but my concern, too, is the harm that may have been done to the baby. As you are preventing neural tube defects, you might be increasing that child's risk of childhood cancer. There's also the danger that because they're taking the folic acid pills, women will be less likely to concern themselves with eating green vegetables during pregnancy. Like everything else in medicine, we often try to look for a pill solution, which has outcomes that we didn't intend, and which doesn't provide the same benefits as a more natural approach.

JOHN ROBBINS: Are there any specific supplements that you tend to recommend for particular groups of people?

DR. JOEL FUHRMAN: I do recommend that some groups of people pay more attention to certain nutrients than others. But I want them to achieve that not by taking the conventional multivitamin, which could have isolated beta-carotene, Vitamin A, folic acid, too much Vitamin E, or one Vitamin E fragment in it. In other words there are negative effects from most multivitamins. Most of the studies done on them show overall minimal to no life span enhancement, because people are mixing things that are potentially valuable with things that are potentially hurtful.

A person on a vegan, vegetarian, or nutritarian diet may not be getting adequate B12. The risk of B12 deficiency goes up with aging as the ability to assimilate B12 goes down. The same might be true with Zinc and with Iodine. It might be beneficial to supplement with Iodine, especially if you're not eating salt in your diet, to make sure you have some presence of Iodine that might not be sufficient in your food. Certain fatty acids are not high in a vegetarian diet, like EPA and DHA. Some people think a person who eats flax seeds and walnuts might make enough. But my twenty years of medical experience drawing blood tests and seeing people develop problems suggests that as people, and especially males, get older, their ability to convert the short chain omega-3 fatty acids into the long chain ones diminishes. To make sure that we're not gambling with people's health, I want to err on the side of safety and confirm that they have an adequate amount of these nutrients present.

JOHN ROBBINS: Your show, *3 Steps to Incredible Health!*, was PBS's top fundraising program for 2011. Why do you think there was such a huge response?

DR. JOEL FUHRMAN: I think it's because people can tell that I'm genuine and that I have credibility that comes from real life experience.

So many people have made recoveries as they have reversed heart disease, reversed diabetes, and gotten rid of asthma, migraines, headaches, and high blood pressure. People watching the show say, "Wow I'm suffering with all kinds of conditions and I'm heading in the wrong direction. But I can be healthy." A lot of people are realizing that they can take control of their health.

Steps You Can Take: Nutritious, Delicious, and Affordable

Get High Nutrient Bang for Your Buck

A nutritarian seeks to get the maximal ratio of micronutrients to calories. For Dr. Joel Fuhrman's micronutrient chart, visit *www.drfuhrman.com*. If you're on a budget, remember that some of the most affordable and nutritionally potent vegetables often include cabbage, carrots, and onions.

Go with Homemade

Americans today spend about 49 percent of their food budget on eating out at restaurants. Even if they're trying to do the right thing, restaurants are under constant pressure to cut costs anywhere they can. This often means lower quality ingredients than you would ever use at home. It also means that in some restaurants, your food is prepared by people who are sick. When you make food yourself, you know what's in it—and you can save a lot of money, too.

Use Shopping Lists—and Stick with Them

Maintain a shopping list, and conduct a quick inventory of your kitchen before you shop to see if you're missing anything important. By thinking your shopping through in advance, you're more likely to get what you actually need, and less likely to be seduced into impulse buying that you'll later regret. There are even shopping list apps, some of which can be shared with a spouse or housemate so either of you can add new things to the list, and you can make sure you don't both shop for the same thing.

Eat Before Shopping

Grocery stores know the power of delectable smells. Everything looks good when your stomach is screaming "feed me!," and that can lead to more impulse buying of things that provide momentary pleasure but really aren't good for you.

Cook in Quantity

Whether you live alone or are part of a big family, making big sauces, pots of soup, casseroles, and other meals saves time in the long run. You can freeze extras for convenient, instant meals, or create meal-sharing arrangements with friends or co-workers.

Take Advantage of Green Polka Dot Box

Green Polka Dot Box is a nationwide (in the United States) natural and organic buyers collective that makes healthy and GMO-free foods available for low prices, delivered straight to the customer's door. Americans can also screen out things like gluten, or animal products, to make it easier to find just what they want. It's not as sustainable as supporting a farmers' market or a local business, but it offers great convenience and pricing, and many quality products. Get more info or sign up at: *www.greenpolkadotbox.com/ foodrevolution*.

Resources for Health and Nutrition

The Ornish Spectrum

www.ornishspectrum.com

Dr. Dean Ornish, M.D., conducted studies that proved, for the first time, that lifestyle changes can prevent and reverse heart disease. The Ornish Spectrum website offers a program that helps you make healthy, sustainable lifestyle changes in what you eat, how much activity you have, how you respond to stress, and how much love and support you have; as well as the background research that underpins the program. Nutrition guidelines, recipes, supplements, and food label guidelines are also available.

Prevent and Reverse Heart Disease

www.heartattackproof.com

Dr. Caldwell B. Esselstyn, Jr., M.D., is an accomplished physician and author of *Prevent and Reverse Heart Disease*, which discusses his groundbreaking program and details his twenty-year study that proves changes in diet and nutrition can actually cure heart disease. The website offers articles, a Q&A section, useful links, and upcoming speaking dates.

Physicians Committee for Responsible Medicine

www.pcrm.org

The Physicians Committee for Responsible Medicine (PCRM), led by Dr. Neal Barnard, promotes preventative medicine through innovative programs. The committee includes physicians, health-care professionals, veterinarians, and compassionate laypersons, all of whom support the mission to advocate for ethical research and promote life-saving nutrition policies and practices. The PCRM website offers breaking medical news, information about diabetes and cancer, educational resources and classes, tools for taking action, blogs, videos and nutritional information, resources for media and citizen organizing, clinical research, legislative focus, and a vegetarian starter kit.

The T. Colin Campbell Foundation

www.tcolincampbell.org

T. Colin Campbell, Ph.D., has been on the forefront of nutrition research for more than forty years. His legacy, The China Project, is the most comprehensive study of health and nutrition ever conducted. The T. Colin Campbell Foundation is a nonprofit organization whose motto is "Scientific Integrity for Optimal Health." The foundation's primary focus is on providing public education about the health benefits of a whole food, plant-based diet. The website organizes medical information by disease, and also by articles written by physicians.

Dr. Joel Fuhrman: How to Live, for Life

www.drfuhrman.com

Dr. Joel Fuhrman, M.D., specializes in preventing and reversing disease through nutritional and natural methods. He is a speaker, as well as the author of seven books, including the *New York Times* bestseller *Eat to Live: The Amazing Nutrient-Rich Program for Fast and Sustained Weight Loss*. Dr. Fuhrman's website promotes his researched ideas about nutrition and health, offers deeper insight into his books, discusses the foundation behind weight loss and the ability to reverse disease, and offers access to resources, a vitamin advisory site, membership, and more.

Mercola.com: Take Control of Your Health!

www.mercola.com

Dr. Joseph Mercola, O.D., founded what has become the most popular natural health site on the Internet. He offers breakthrough (and sometimes controversial) insights on many aspects of health and natural living. Visit *www.cookware.mercola.com* for more on his popular ceramic cookware.

Dr. McDougall's Health and Medical Center

www.drmcdougall.com

Dr. John McDougall, M.D., is a physician, nutrition expert, and author who teaches better health through vegetarian cuisine. Dr. McDougall is the founder and medical director of the nationally renowned McDougall Program, a ten-day, residential program located at a luxury resort in Santa Rosa, California. Dr. McDougall's website provides short video lessons, hot topics, information about the Health Center live-in program, free lectures, the McDougall Diet, and an Adventure Travel page.

TeraWarner.com: Fresh Thinking for Healthy Living

www.terawarner.com

Tera Warner is the founder of the world's largest online resource of raw food cleansing and detoxification programs for women. Her website offers information and support for her programs, including the 3-Day Green Smoothie Challenge, the 10-Day Juice Cleanse, and 7 days on a "rabbit food" program. The site hosts summits and a Women's Wellness University that offers information on various health and wellness topics, a blog, recipes, and online courses.

Brenda Davis, Registered Dietician

www.brendadavisrd.com

Brenda Davis is a registered dietitian and nutritionist, a leader in her field, and an internationally acclaimed speaker and author of seven books. Brenda's website offers information about Brenda, her books, her writings, and her schedule, as well as a variety of resources to assist in making choices that promote and sustain health and well-being.

NutritionFacts.org

www.nutritionfacts.org

NutritionFacts.org is a nonprofit, charitable organization founded by the Jesse and Julie Rasch Foundation in partnership with Michael Greger, M.D. Dr. Greger scours the world of nutrition-related research, as published in scientific journals, and presents that information in short, easy-to-understand video segments. Journal articles are also available.

Institute for Integrative Nutrition

www.integrativenutrition.com

Since 1992, Integrative Nutrition has been the cutting-edge leader in holistic nutrition education. As the largest nutrition school in the world, they seek to educate and transform the lives of their students with training from the world's foremost experts in nutrition and wellness. Their curriculum teaches a wide variety of skills in health coaching, nutrition education, business management, and healthy lifestyle choices.

PART II

GMOs: How Dangerous Are They?

Did you know that more than 70 percent of the foods in American restaurants and supermarkets contain genetically modified organisms (GMOs)?

Is there a link between the spread of GMOs and the recent dramatic increase in food allergies?

There are disturbing reports linking genetically engineered foods to toxic and allergic reactions in people; to sickness, sterility, and fatalities in livestock; and to damage to virtually every organ studied in lab animals. Should you be alarmed?

Why is Monsanto spending millions trying to stop GMOs from being labeled, even though 90 percent of the American public supports labeling? If their products are safe, why are they so afraid of us finding out what we're eating?

6

Jeffrey Smith

Take Genetically Engineered Foods Out of Your Diet— Immediately!

Jeffrey Smith is an international bestselling author and filmmaker, and a leading spokesperson on the dangers of Genetically Modified Organisms (GMOs). He has counseled world leaders from every continent, changed the course of government policies, and is now orchestrating a tipping point towards consumer rejection of GMOs through his programs at the Institute for Responsible Technology. Jeffrey's book, Seeds of Deception, *is the world's bestseller on GMOs, and his film,* Genetic Roulette, *exposes serious health risks of the Genetically Modified (GM) foods Americans eat every day.*

Jeffrey delivers the alarming news that GMOs are leading to a massive increase in human exposure both to the Roundup herbicide, and to the Bt toxin that has been genetically engineered into every cell of corn and other plants. Jeffrey gives you the truth that everyone deserves, and more importantly, he tells you what you can do.

JOHN ROBBINS: Monsanto and its allies continue to fight ferociously every attempt to require labeling of foods made from genetically engineered seeds. They say there is no reason for the public to be concerned. But many people are in fact deeply concerned about the potential dangers. Do they have reason to be?

JEFFREY SMITH: Yes, they certainly do. The American Academy of Environmental Medicine reviewed the animal feeding studies that have been published in peer-reviewed journals and found that the animals that were fed the same GMOs that are in our diet had reproductive disorders, immune system problems, gastrointestinal problems, accelerated aging, organ damage, and dysfunctional regulation of cholesterol and insulin. On top of that, we have story after story of animals—livestock as well as pets—that have been taken from a GMO diet and put onto a non-GMO diet and their behavior improves. The death rate, stillborn rate, and use of medication go down, conception rate, and litter size go up, and even behavior improves. Now we have many doctors saying that when the patients avoid GMOs, a lot of their symptoms go away as well.

JOHN ROBBINS: We see GMOs linked to toxic and allergic reactions in humans, to sickness, sterility, and fatalities in livestock, and to damage to virtually every organ studied in lab animals. And yet the biotech industry continues to say that the health fears have been disproven. I heard one industry representative say recently that a trillion GMO meals have been served and not a single case of GMO-induced illness has turned up. How do you respond to that kind of statement?

JEFFREY SMITH: Well this is one of the most unscientific statements that is made by the biotech proponents. First of all, we do know of a genetically modified food supplement in the 1980s that was directly responsible for the deaths of about a hundred Americans and caused five to ten thousand to fall sick or become permanently disabled. It was L-tryptophan, an amino acid used for stress and insomnia, produced from genetically engineered bacteria. Only the brand that was genetically engineered was responsible for this deadly epidemic. What was telling was that it took four years to discover that the epidemic was actually taking place. It took four years even though the symptoms were screaming to be discovered. The disease was new, acute, and fast-acting. If the GMO crops on the market are simply contributing to an increased rate of existing disease like cancer or heart disease or diabetes, or if they are producing nonacute reactions or reactions that take place over many years or generations, we would never be able to identify GMOs as the cause. So the statement that GMOs have not caused any problem is false. What we do know is that they haven't been looking for it, and that there is no way to identify it because of an utter lack of post-marketing surveillance, human clinical trials, and long-term animal feeding studies.

JOHN ROBBINS: It sometimes seems to me that not only are we not looking for it, but also that the system has been set up to make it harder for anyone who wanted to look to actually be able to identify and isolate any difficulties that were ensuing.

JEFFREY SMITH: That's true for a number of reasons. First of all, the scientists that want to do research are often denied access to the patented seeds by the companies that own them. Second, we have documented over and over again that when scientists do discover problems, they are often fired, stripped of responsibilities, threatened, forced out, denied funding, or denied tenure. For example, the Russian Academy of Sciences' senior researcher Irina Ermakova discovered that when she fed female rats genetically modified soy, more than half of their babies died within three weeks. They were sterile and were also quite a bit smaller. She told me that her boss, under pressure from his boss, forbade her from doing any further GMO research. There were methods used to try to intimidate her, documents stolen from her laboratory and burnt on her desk, and samples stolen. One of her colleagues tried to comfort her by saying "well maybe the GM soy will solve the overpopulation problem." She wasn't impressed. She is just one of numerous scientists who have been basically silenced or, in the case of Andrés Carrasco, physically attacked. An organized mob of more than a hundred people attacked him and his friends when he tried to give a talk on the birth defect links to Roundup, Monsanto's herbicide that is used in conjunction with Roundup Ready crops.

JOHN ROBBINS: Many of Monsanto's GMO crops have been engineered to withstand massive sprayings of Roundup, the company's proprietary herbicide. I have seen ads where Monsanto claims Roundup is biodegradable and it is safe. What is the actual environmental and health impact of Roundup?

JEFFREY SMITH: Well Monsanto got convicted of false advertising by a New York Court when they used to say that Roundup was biodegradable. But that didn't stop them from saying the same thing in Europe until the French court nailed them and forced them to pay a fine for false advertising. Roundup, it turns out, is linked not only to birth defects but also cancer, endocrine disruption, smaller sperm counts, abnormal sperm, and a host of other disorders. It was originally patented as a broad spectrum chelator, which means it binds with nutrients and does not make them available to plants. When it is sprayed on plants it deprives those plants of nutrients, making them weaker. Then it promotes pathogens in the soil, which kill the plant. So it creates a perfect storm of disease and death.

The largest proportion of the diet of livestock in the United States is Roundup Ready crops—specifically soy, corn, cottonseed meal, canola meal, sugar beet pulp, and alfalfa. Now that they are eating nutrient-deficient crops, there is a universal deficiency of certain nutrients in the livestock in the United States. When we eat these sick and nutrient-deficient animals, and when we eat the nutrient-deficient Roundup Ready crops directly, how does that affect our own nutrient balance? But there's more. When we eat crops that have higher levels of Roundup, we absorb the Roundup, which not only has direct toxicity, but also can chelate, or bind with, nutrients in our body, depriving us of those very, very important nutrients.

Roundup is now found in air samples, rain samples, water samples, and in our urine. It is found in our blood and in the blood of fetuses. It is dangerous in very minute concentrations. There was some recent evidence that Roundup kills beneficial gut bacteria. In fact, within cows it kills the gut bacteria that keeps down the population of botulism and so there is supposedly an upsurge of botulism now in cows, and possibly also in the human population, linked to the all-pervasive Roundup in our diets.

JOHN ROBBINS: As bad as Roundup is, I am seeing something worse on the horizon. As use of Roundup has been increasing, it has catalyzed an increase in herbicide-tolerant weeds, thus forcing farmers to switch to other chemicals including the acutely toxic 2,4-D. Most of us are familiar with this as the primary ingredient in Agent Orange, the defoliant used in Vietnam that has been linked to thousands of deaths and birth defects. It seems like something out of a bad science fiction movie, but is it actually true that Dow Chemical wants to solve the problem by creating corn and soybeans that are resistant to 2,4-D, which would lead to millions of acres of U.S. farmland being sprayed with a chemical warfare agent?

JEFFREY SMITH: You are absolutely right, it is stranger than fiction and this is their response to the inevitable emergence of herbicide-tolerant weeds. Roundup Ready weeds have emerged on more than 13 million acres in the United States. Instead of going back to more sustainable programs and approaches, they are introducing these Agent Orange crops, which are guaranteed to flood our bodies and our environment with an acutely toxic substance. It is an example of the kind of skewed and profit-driven thinking of the industrial model of agriculture that completely ignores the consequences for health, environment, wildlife, and future generations.

The introduction of GMOs on its own right, even without its associated chemicals from these herbicide-tolerant crops, is pretty scary stuff. We are putting genes into the environment, and we have no technology to be able to recall them. The genes already released become self-propagating pollution in the gene pool, and can outlast the effects of global warming and nuclear waste. On top of that, they affect everyone who eats, so the exposure and the risk of GMOs to all living beings and all future generations is unprecedented in our history.

JOHN ROBBINS: Since the introduction of widespread use of GMOs, we have seen a substantial increase in food allergy rates, particularly in kids. Do you see a connection, and if so, what do you base that on?

JEFFREY SMITH: I definitely see a connection. Soon after GM soy was introduced to the UK, soy allergies skyrocketed by 50 percent. It was not followed up to see if GM soy was the cause, but there are many reasons why it could have been. For example, there are high residues of Roundup in the soy, which could lead to intolerance. There is as much as a seven-fold increase in a known allergen in soy called trypsin inhibitor that might cause the alarming increase. There was also a new allergen found in the GM soy compared to a non-GM wild type of soy, and that might cause the problem.

When mice were fed GM soy, there was damage to their pancreatic cells, which caused a reduction in the production of digestive hormones or digestive aids. If it takes longer to break down proteins, then that means the person will have more opportunity to experience an allergic reaction. Even the Roundup Ready protein has properties of a dust mite allergen, so if you are allergic to dust you might be allergic to Roundup Ready soybeans.

JOHN ROBBINS: Most GMO crops on the market today have been engineered to either be resistant to Roundup, and/or to produce the Bt insecticide in every cell of the plant. Do you see a possible link between the Bt crops and any health dangers?

JEFFREY SMITH: Unfortunately, yes. The Bt crops have been designed to produce their own insecticide, Bt, that breaks open the stomach of insects and kills them. The Bt toxin comes from soil bacteria and it has been used in farming for a long time as a spray. But the genetic engineers take the gene that produces the toxin and insert it into corn and cotton plants. The Bt toxin is known in its natural state to create immune responses and allergic-type responses. When it was sprayed for Gypsy Moth infestation in the

Pacific Northwest, about 500 people had allergy or flulike symptoms, and some had the go to the hospital. When it was fed to mice, they had immune system responses.

Now mice fed the Bt corn are getting massive immune responses. Farm workers in India by the thousands are getting allergic reactions and rashes all over their bodies from just touching the Bt cotton, and we see that animals grazing on the Bt cotton after harvest in India are dying by the thousands.

In addition to the presence of the Bt toxin in massive quantities, there are other reasons why Bt crops may be triggering food allergies. The process of insertion of the gene and the cloning of the cell into a plant causes massive collateral damage. This can result in hundreds of thousands of mutations up and down the DNA, and changes in gene expression by hundreds of genes. In Monsanto's Bt corn, a silenced gene was inadvertently turned on, producing a known allergen in the corn. This was only discovered after the corn was on the market, and did not result in any warnings to those who might be reactive.

There is also evidence that by increasing the allergic reactions to one item, you may end up being sensitized to other formerly harmless foods. So exposure to GM corn or soy might also increase reactivity to peanuts and dairy, for example.

JOHN ROBBINS: If someone eats Bt corn, for example, is it possible for the gene that codes for the production of Bt to somehow transfer to gut bacteria in the human intestine?

JEFFREY SMITH: Actually this is the subject of the only human-feeding study ever published, and it wasn't on Bt, it was on the Roundup Ready soybean. They found that in fact part of that Roundup Ready gene did transfer to the bacteria living inside human intestines and that these folks had Roundup Ready gut bacteria. This suggested that the gene, once transferred, continued to function. Now they didn't follow up to see if the Bt toxin gene also transfers, but this is a critical question. If the toxin-producing gene is in corn chips, for example, and if it then transfers to our gut bacteria, then our intestinal flora become living pesticide factories, producing the Bt toxin over and over again.

We know that the Bt toxin might be an allergen, but it gets worse. The reason it kills insects is that it breaks open their stomach. It pokes holes in the cell walls of their digestive track. Now, according to the biotech industry, this should only happen to insects and not to humans. That was until 2012 research published in *The Journal of Applied Toxicology* showed that it happens also in humans—that it pokes holes in the cell membranes, creating pores in the human cells and causing leakiness. Is it possible that eating the Bt toxin, especially if it turns our intestinal flora into living pesticide factories, could be eroding the lining of our intestinal walls and our stomachs? According to numerous medical reports and scientists, if we have permeable intestines, then undigested food particles will make their way into the bloodstream, which can trigger allergic reactions, autoimmune disease, inflammation, and may even be linked to autism.

The biotech industry claims that the Bt toxin is destroyed in the digestive process, but that was disproven in 2011 when a study came out in Canada showing that 93 percent of pregnant women had the Bt toxin in their blood. In this study, it was also found in the blood of 80 percent of their unborn fetuses. Now this may be a direct result of consuming foods that contain the Bt toxin, or it could be because the gut bacteria of these humans is now producing the Bt toxin on a regular basis. If we have colonized the gut bacteria of North Americans with Bt toxin-producing genes, that might explain the huge rise in gastrointestinal problems, including the 40 percent increase in inflammatory bowel disease since GMOs were introduced. It could explain the multiple chronic illnesses that rose from 7 percent to 13 percent in the first nine years of GMOs, the increase in food allergies, in autism, and in a whole spectrum of other diseases. This would be a fundamental problem that could affect the immune system, the digestive system, the development of babies, and so much more.

JOHN ROBBINS: The risks that we are taking are obviously immense and the justification that Monsanto and its allies have been proposing for years is that we need these crops to feed the world. They have been promising us crops that would grow in drought conditions and in saline soils, crops that would be nutritionally superior and that would act as medicines. They have been promising us increased yields, and that their crops would lower the use of pesticides. Almost none of that has actually occurred. It would be wonderful if in fact this technology could help us feed our growing numbers. But there seems to be no evidence that it is doing that.

For example, in 2011, the USDA approved Monsanto's first-ever supposedly drought-resistant crop. However, according to the USDA report, this genetically modified corn only works under conditions of moderate drought. We already have conventional corn varieties that substantially outperform this GMO corn in serious drought conditions. So it seems to me that Monsanto's greatly hyped and new supposedly drought-resistant corn is actually no improvement at all.

JEFFREY SMITH: Absolutely. In fact, this is just another in a long string of promises that they have been using as an excuse for their technology since it was introduced. I talked to a former Monsanto representative, Kirk Azevedo, who was recruited into Monsanto and agreed to join them only after reading the words of then-CEO Robert Shapiro who talked about these glowing promises. When Kirk went to the employee orientation meeting at Monsanto's headquarters in St. Louis, he got up and described how excited he was, and he re-quoted the words of Robert Shapiro about the company's vision for helping the world. After the meeting a Vice President pulled him aside and said: "Wait a minute. What Robert Shapiro says is one thing. What we do is something else. He is the front man that tells a story. We don't even know what he is talking about. We are here to make money."

The experts in feeding the world also agree that GMOs currently have nothing to offer that helps us to feed the world, eradicate poverty, or create sustainable agriculture. This was made clear in the comprehensive *International Assessment of Agricultural Knowledge, Science and Technology for Development* report on world agriculture, signed off on by fifty-nine nations. Credible researchers know very well that the supposed social benefits of genetically engineering our food crops have so far been just a PR line.

JOHN ROBBINS: One of the most alarming environmental developments of our time is Colony Collapse Disorder. Commercial beekeepers in the United States are saying their industry is on the verge of collapse. The food crops that are dependent on bees for pollination include most of our fruits and many of our vegetables. There is now increasing evidence that nicotine derived insecticides called neonicotinoids, marketed by the German agrichemical giant Bayer, are the primary culprits. They use these to coat the seeds prior to planting and they use higher concentrations of these poisons in the GMO seeds. Is there a reason that the GMO seeds require more of the insecticide coating than the normal seeds? And do you think there could be a connection between GMOs and Colony Collapse Disorder?

JEFFREY SMITH: GMO corn has been genetically engineered to produce the Bt toxin, which kills the corn root worm. When they tested it they found that the amount of Bt that the GMO corn produced in its earliest stages of germination and growth was insufficient to protect the roots, which left the plant vulnerable for probably a few weeks. So for these particular plants, they coat the seeds in neonicotinoids using a time-released program. The neonicotinoids go into the plants themselves and release over several weeks. They put a higher concentration of neonicotinoids in the GMO corn seeds—up to five times the amount—so that the infant plant is protected until it has matured enough for the Bt toxin to take over protection of the plant.

JOHN ROBBINS: So if it is taken into the seed, it goes into the land. Is it also going to be expressed in the pollen that the bees collect?

JEFFREY SMITH: Yes, but it might hurt the bees even before the pollen is created. When farmers deliver the seeds and plant them, there is a dust residue that ends up on the other crops that the bees will pollinate, including daffodil or dandelions. They have been able to measure levels of this dust borne insecticide in the bees that have died and also in their hives. Then later, as the plant gets older, the pollen itself also carries this insecticide. So whether it is in the dust or in the pollen, there is a perfect storm for increasing the exposure. What happens is the bees can either die directly, or they can lose the ability to find their way back to the hive, which is one of the signature characteristics of Colony

Collapse Disorder. The bees just can't figure out a way to get home, and they die on the ground.

JOHN ROBBINS: Is there any data to support the theory that, in addition to use of neonicotinoids with GM crops, there could be a link between the Bt toxin and Colony Collapse Disorder?

JEFFREY SMITH: There are higher levels of Colony Collapse Disorder in the United States compared to other countries that have lower GMO planting rates. That could be a result of the higher use of neonicotinoids with GM corn, but it could also be exacerbated by the direct impact of bees gathering pollen from fields growing genetically modified corn.

We have two data points to support this. One, in Germany, showed that when bees gathered genetically modified pollen, the genetically modified genes transferred into the DNA of the microorganisms living inside the guts of the bees, just as it happens in humans. Another study showed that when bees gathered pollen from corn that was genetically engineered to create the Bt toxin, these bees ended up succumbing to higher rates of a viral infection. The control bees, whose pollen was not genetically modified, did not. This would suggest that the Bt toxin, although it wouldn't kill the bees directly, might compromise their immune system, allowing them to be more susceptible to diseases and disorders. Then there is the fact that many bees are fed high-fructose corn syrup during the winter because big honey operations take all of the honey that would nourish bees through the winter. That corn syrup is mostly GMO, so that is another way that GMOs could be impacting and harming the health of the bees. We don't have any smoking gun on this. I think the nicotine-derived pesticides are the likely leading cause. But GMOs could be a supporting player in this high-stakes drama.

JOHN ROBBINS: Jeffrey, you see the dangers and you articulate them so clearly. How do you live with the knowledge of what we as a culture are doing to our health and to the whole web of life? What keeps you going? How do you retain your strength in the face of the obstacles and the discouragements that are inevitably part of this activism?

JEFFREY SMITH: I am actually very optimistic. We saw what happened in Europe in 1999 when the gag order was lifted on Dr. Árpád Pusztai. He was a researcher evaluating how to test for the safety of GMOs under a UK government grant. His protocols were supposed to be implemented into the EU law for evaluation of subsequent approvals. He discovered quite accidentally that GMOs were unsafe, and he went public with his concerns. He was fired from his job, gagged with threats, and maligned and attacked for seven months. But when his gag order was lifted by an Order of Parliament, the media ran a firestorm of articles about GMOs—more than 700 within a month in England alone. The resulting consumer awareness about GMOs and their possible health impacts was sufficient to create a tipping point of consumer rejection. Within ten weeks after the gag order was lifted, virtually every major food company in England committed to stop using GM ingredients.

We saw what happened as we and others educated U.S. consumers about the health risks of genetically engineered Bovine Growth Hormone and how it is linked to cancer. It has now been kicked out of Wal-Mart, Starbucks, Yoplait, Dannon, and most American dairies.

If you look at the numbers, how many people need to avoid brands that contain GM ingredients before we reach a GMO tipping point? I think that any drop in market share whatsoever that a Kraft Food manager can attribute to the growing anti-GMO sentiment in the United States would be sufficient to cause them to quickly abandon GM ingredients. To get a drop in market share that they can identify as GMO-related, we think we need only about 5 percent of U.S. shoppers avoiding GM ingredients.

Ninety-three percent of Americans polled say they want foods that contain GMOs to be labeled, and significantly, 53 percent of Americans say they wouldn't eat GMOs if they were labeled. Now the major food companies in the United States have already removed GMOs from their European brands because of a tipping point of consumer rejection that occurred there. If they see that Americans are going to remove their brands from their shopping carts over GMOs, then these companies would rather eliminate GMOs than admit that they use them. So the moment that labeling is required is a watershed moment in the struggle.

Even without a labeling law in place, we have so many people trying to avoid GMOs. We have thousands of doctors prescribing non-GMO diets, mothers protecting their kids, and religious groups protecting members on the basis of their religious beliefs.

I have been working on this issue since 1996, and I can see the signs of a coming tipping point. Non-GMO labels are one of the fastest growing label claims in America on food products sold in grocery stores. We are seeing now the expression of growing anti-GMO sentiment, not only in labeling bills and ballot initiatives, but in rallies held at state houses all over the United States. We now have millions of people engaged for the first time. I think we are seeing a food revolution in the area of GMOs like never before, and I think it is just a matter of time. I hope that I am going to put myself out of a job very soon.

7

Dr. Vandana Shiva
How to Stop Big Ag

Vandana Shiva, Ph.D., is a world-renowned environmental leader and thinker. Starting from her mother's cowshed in India, she founded the Research Foundation on Science, Technology, and Ecology. Vandana is also founder of Navdanya ("nine seeds"), a movement promoting diversity and use of native seeds. Vandana is author of many bestselling books, and the recipient of numerous awards and honorary degrees including the Alternative Nobel Peace Prize. In 2011 Forbes *named her "one of the 7 most influential feminists in the world."*

Vandana reflects on the tragic irony that in the land where Gandhi spun freedom through cotton and the spinning wheel, 95 percent of the cotton is now controlled by Monsanto. The result? More than 250,000 farmers in India have committed suicide since Monsanto began to take control of the cotton industry. Dr. Shiva is working with 500,000 farmers in India and worldwide, creating sixty-six seed banks, and mobilizing millions to work for food sovereignty. This is her inspiring story and her call to action.

JOHN ROBBINS: There are corporations that are seeking to profit from and control every aspect of the food-supply chain, from seeds to land to water. What is at risk and what is being threatened by the efforts of Monsanto and other biotech companies to patent the seeds we depend on for our food?

DR. VANDANA SHIVA: I think the very future of life is at risk with patenting and the attempt by a handful of corporations to control our seeds.

First, corporations with their centralized control cannot deal with the diversity that the planet has. That diversity has been nourished by caring farmers at a small-scale level. For example, India evolved 200,000 varieties of rice. Small farms can grow twenty-five to thirty varieties of rice. Navdanya, on its farm in Doon Valley, grows about 700 varieties of rice. But doing that requires a kind of love and care that is only possible at the small scale. We have had Monsanto around for the past fifteen years, commercializing GM crops. What have they given us? Mainly four crops: Roundup Ready canola, Roundup Ready soya, Roundup Ready and Bt corn, and Bt cotton. Four crops with two traits. One trait enables more spraying of herbicides, which are causing more toxins in the plants and creating resistant superweeds. The other trait is Bt crops, which are creating resistant superpests. So we're getting superweeds and superpests, and the loss of diversity.

Second, allowing the patenting of life-forms is eroding the very ethical fabric of our existence on this planet. We are part of the earth family; we are one among many species. If a handful of corporations now claim to be creators and inventors of life on earth, that is going to put our thinking and our relationship with the earth in disarray. In addition, the only reason corporations like Monsanto take a patent is to create a monopoly, giving them the right to extract superprofits in the form of royalties. They get these patents on very false grounds of having invented seeds—which they haven't. They poison the seeds, and they should be treated as polluters and fined, not rewarded with a monopoly.

When the British colonized the Third World, they took over the land and revenues from the land, and dispossessed the Indian people. That is what created the hunger and famines and landlessness in India, or in Africa, or in Latin America, where it was more the Spanish and the Portuguese who colonized. Today the colonization is of the seeds themselves. The small persons who are given the seed, who are the source of the seed, have never before had to buy seed. They don't have money and when the companies' agents come they say "You are going to be a millionaire! You are going to get 1,500 kilograms of cotton per acre. Just sign on this little piece of paper." The signing on the piece of paper is a mortgage on the land.

The farmer thinks, "If I get that much yield, if I get that many returns, I will be able to pay back this loan, so what's the problem?" The loan is against an 8,000 percent increase in seed prices and the false promise of pest control. This is a guaranteed debt trap. Farmers in the cotton belt of India today have lost control over the seed. Ninety-five percent of it is owned by Monsanto, in just a decade! In the land where cotton evolved, in the land where Gandhi spun freedom through cotton and the spinning wheel, today 95 percent of the cottonseed is owned and controlled by Monsanto.

Monsanto's superprofits and royalties have trapped our farmers in debt, and a quarter-million Indian farmers, mostly in the cotton areas, have committed suicide. Patents on seed mean genocide.

JOHN ROBBINS: What are the economics at play that are forcing people to have no possible recourse?

DR. VANDANA SHIVA: I think it is a mixture of factors. Monsanto actually emerged as a war chemicals industry. It is known for Agent Orange, and for toxics. It wasn't ever in seed and agriculture. This is a recent entry, because they realized controlling the seed means controlling the entire food chain and the profits they can make from that are so much more than they can make at any other level. So in a way they brought war to our farmlands. They brought war against our farmers. The economics of it are first and foremost the economics of a monopoly, created by a highly undemocratic, international trade treaty, which brought clauses on control over the seed.

There is an article 27.3(b) in the Trade Related Intellectual Property Rights Agreement, which is the agreement of the World Trade Organization that governs seeds, agriculture, patenting, and intellectual property. Monsanto is on record saying, "We were the patient, diagnostician, and physician, all in one in drafting this treaty. We defined the problem." And the problem they defined was that farmers save seeds and the solution they offered was that if farmers saved their seeds, it should be treated as an intellectual property crime. Worse, Monsanto has gone further to say that if Monsanto contaminates a farmer's field with genetic pollution from their seeds pollinating, the contamination represents not a pollution for which Monsanto should be liable, but rather a theft of Monsanto's technology.

This is what they did with Percy Schmeiser, a Canadian farmer whom they sued for $200,000 after they contaminated his canola crop. That is why I joined the group of eighty-four plaintiffs who brought a case to the District Court in New York to stop Monsanto from using patent law to sue organic farmers after contaminating their crops. Unfortunately the District Court dismissed the case, saying those farmers have no standing. You have no standing to protect your crops? You have no standing to protect your rights as farmers? This is the ultimate abuse of human rights.

Another economic reality is that Monsanto and their agenda really has taken over governments. It has taken over the White House, it has taken over the USDA, and it has even taken over our Indian Prime Minister's office. Our entire seed and agriculture policy is being drafted by these companies. The United States passed the Global Food Security Act of 2009 with the singular focus of distributing GM seeds around the world. It was a 7.7 billion dollar subsidy to Monsanto using the arm twisting of USAID.

Economics is no longer about fair play. It is no longer about competition. It has come to be about stealing of governments, stealing democracy, stealing our seeds, and stealing our freedom. This is the ultimate test, which is why I have dedicated my entire life to fighting it and have created an entire organization and movement.

I am not the kind to build institutions. I wanted to be a free bird doing physics quietly with my little equations. That was my dream of life. But instead I am working with 500,000 farmers, mobilizing millions, creating real alternatives with sixty-six community seed banks, and training more than 300 farmers every month to be free in their seed and free in their food. It's because there is too much at stake. In fact, everything is at stake.

JOHN ROBBINS: Shouldn't it be the responsibility of Monsanto and of farmers who plant GMO crops to ensure that genetically engineered DNA does not trespass onto neighboring farmlands?

DR. VANDANA SHIVA: Yes, it should. There is now a United Nations International Treaty on Biosafety. I had a big role in introducing the clauses that led to it at the Earth Summit in 1990 in Rio. Under this treaty, there is a new protocol on liability. We need to see that every country implements this protocol so that Monsanto is held responsible for the contamination it is causing.

JOHN ROBBINS: After Monsanto introduced Roundup Ready canola, organic canola became virtually nonexistent in its pure form because of the contamination. We are starting to see a similar pattern now with corn, soybeans, cotton, sugar beets, and alfalfa. Is it even possible for transgenic seed to coexist with organic seed, or will it eventually contaminate the entire organic world?

DR. VANDANA SHIVA: I don't think coexistence between GM crops and organic is at all possible. Any official who says it is, is lying through their teeth. Last year at a conference organized by Prince Charles in Washington, I had been invited to speak at the International Session. Tom Vilsack, the U.S. Secretary of Agriculture, began his speech by saying "I have two sons: GMO and organic, and I love them both and I want them to coexist," and someone got up from the audience and said: "We're very sorry, but one of your sons is a bully and won't allow the other one to exist." And that is the reality.

Genetic engineering is done in the lab, and then the plants are released in the fields and they behave like any other plant: they cross-pollinate. The pollen spreads by wind, and genetic contamination is inevitable. This is why we've said to stop the deliberate release. Because the systems that should build a caution around this technology have all been deliberately dismantled.

At the Rio Earth Summit in 1992, when we got the biosafety clauses to include the Convention on Biological Diversity, U.S. President Bush stormed out and refused to sign the convention. He came back and immediately asked his Vice President, Dan Quayle, to advise him. The Vice President became an advocate for what was called substantial equivalence. They just declared, "Treat every genetically modified crop as equivalent to the crops from which you robbed the parent material." This is a policy of don't look, don't see. But this is also a policy of what I have called ontological schizophrenia.

When you want to patent that same seed, you say, "I made something totally new, it didn't exist in nature, no farmer has ever bred it, I am the inventor, give me a patent." And then when we turn around and say, "And this will have consequences so you should own the contamination and therefore pay for it and compensate the farmers whose fields you have destroyed," they say "No, no, it's just like nature made it. This is a natural organism." You can't have the same organism being natural when it comes to responsibility for pollution, and totally an invention of Monsanto when it comes to taking rights over the seed. The whole structure is so dishonest and that is why I am starting a huge campaign on seed freedom. We want to remind the world that there is no food sovereignty without seed sovereignty, and there is no food freedom without seed freedom.

JOHN ROBBINS: My hope is that one seed at a time, one farm at a time, one meal at a time, we can break free from corporate food dictatorships, and in the process, create a vibrant and fertile food democracy.

Steps You Can Take: Protecting Your Family from GMOs

Get the Non-GMO Shopping Guide

The Non-GMO Shopping Guide features hundreds of brands that are currently enrolled in the Non-GMO Project. This purse/pocket-sized guide will help you identify and avoid foods that contain genetically modified organisms (GMOs) while you shop. Available in English and Spanish at *www.nongmoshoppingguide.com.*

Get the Non-GMO Mobile App

The Center for Food Safety created a mobile app to help you find and avoid genetically engineered (GE) ingredients wherever you shop. This guide gives you valuable information on common GE ingredients, brands to look for and look out for, and common-sense tips to keep you in the know. Available at *www.foodrevolution.org/gmopack.html.*

Be a GMO Buster

Next time you are shopping for food, take five minutes to read labels and see how many products you can identify that contain GMOs. For a list of products that are likely to contain GMOs, visit: *www.responsibletechnology.org/gmo-basics/gmos-in-food.*

GMO Empowerment Resources

Institute for Responsible Technology

www.responsibletechnology.org

Founded by Jeffery Smith, The Institute for Responsible Technology (IRT) is a world leader in educating policy makers and the public about genetically modified foods and crops. The IRT website is a comprehensive repository of reliable and current resources regarding GMOs by offering online videos, podcasts, blogs, and reports, many of which are free. Their newsletter is a trusted source of updates, events, and news for subscribers.

Navdanya

www.navdanya.org

Founded by Dr. Vandana Shiva, Navdanya is a network of seed keepers and organic producers spread across sixteen states in India. The website offers information about seed, food, and land sovereignty; sustainable agriculture; water democracy; climate change; organic food production; organic certification; and organic shopping lists. There are also links to several campaigns that come from a global and/or India-based perspective.

The Center for Food Safety

www.centerforfoodsafety.org

The Center for Food Safety (CFS) is a nonprofit, public interest and environmental advocacy membership organization established in 1997 for the purpose of challenging harmful food production technologies and promoting sustainable alternatives. CFS combines multiple tools and strategies in pursuing its goals, including litigation and legal petitions for rule making; legal support for sustainable agriculture and food safety constituencies; as well as public education, grassroots organizing, and media outreach.

The True Foods Network

www.truefoodnow.org

The True Foods Network is the Center for Food Safety's grassroots action network, with more than 180,000 members across the United States. The True Food Network is where concerned citizens can voice their opinions about critical food safety issues, and advocate for a socially just, democratic, and sustainable food system. Their website contains information and links to their sponsored campaigns, a non-GMO shopper's guide and smart phone app, publications, and a Take Action page.

The Non-GMO Project

www.nongmoproject.org

The Non-GMO Project is a nonprofit organization that offers North America's only third-party verification and labeling for non-GMO foods and products.

PART III

Eating as if the Earth Mattered (Which It Most Certainly Does!)

We face some big challenges on planet Earth. Hunger is now taking the lives of 17,000 children daily.

The rainforests are being slashed, burned, and turned into deserts by cattle ranching and large-scale soy plantations.

Perhaps most disturbing of all, leading scientists believe our global climate is being dangerously destabilized. That means we all may face more droughts, floods, hurricanes, and coastal flooding.

You can be part of the solution.

If you want to think globally and act locally, you can't get more local than the food on your plate.

What you eat has a bigger impact on our world than you probably ever imagined.

Do you want a world where everyone has enough to eat?

Do you want a world with a stable climate and healthy ecosystems?

It turns out you can help change the world with your knife and fork. Deliciously.

Good food isn't just healthier and better for the world—it's good for your conscience, too.

8

Bill McKibben

The Greatest Threat to the Survival of Civilization as We Know It

Bill McKibben is the founder of 350.org, which has become the biggest grassroots climate campaign on the planet. Described by The Boston Globe *as "probably the most important environmentalist" in the United States, Bill literally wrote the first popular book on global warming,* The End of Nature. *And he has now authored another dozen powerful books including* Earth: Making a Life on a Tough New Planet.

Why does Bill think climate change might be the greatest threat to the survival of the modern world? If this concerns you, chew on this: Cows impact our climate more than cars. What's the most climate-friendly diet?

JOHN ROBBINS: You have been fighting global warming now for nearly a quarter century. During this time, as you have often pointed out, both the scale and the pace of the peril have increased substantially. Yet our politicians have mostly ignored the increasingly overwhelming scientific consensus. What kind of future are we entering?

BILL MCKIBBEN: When I wrote *The End of Nature* in 1989, we knew most of what we needed to know about climate change. We knew that when you burned coal, gas, and oil, you put carbon in the atmosphere. We knew that the molecular structure of carbon trapped heat that would otherwise radiate back out to space. The only thing we didn't know was how hard and fast this was going to pinch. The story of the past twenty years is that it is pinching harder and faster than we would have expected. So far we have raised the temperature of the earth about 1 degree. We are trapping about three-quarters of a watt of extra solar energy per square meter of the earth's surface. These numbers are not huge, and we wouldn't have thought twenty years ago that they would lead to huge results, but the earth was more finely balanced than we realized.

We are moving into a new and profoundly altered future. There is 40 percent less sea ice in the Arctic in the summer; the oceans are about 30 percent more acid; and the atmosphere holds a staggering amount more water, which is leading to more drought and more floods. We are at the beginning of the climate change era. There is much worse to come unless we get our act together quickly, but already the damage is far greater than we would have imagined twenty years ago.

The best science would seem to indicate that while we can no longer stop global warming, we still have a small window in order to prevent it from getting completely out of control. That small window is closing and that is why I believe that we are in an emergency.

JOHN ROBBINS: Nature is immensely powerful. It seems that we are unleashing potentially unstoppable planetary forces.

BILL MCKIBBEN: Yes, we are unleashing very strong forces like droughts and floods. The expectation is that rapid climate change will drive to extinction a huge percentage of the species on this planet—perhaps the same kind of percentage that we have seen in past geological epochs, only when huge asteroids struck the earth. In this case, of course, the asteroid is us, and the frustrating part is that we don't need to do it. We know much of what we need to know to avert this. We are just not doing it, because it is in the strong financial interest of a small group of human beings to keep us on our present course.

JOHN ROBBINS: With so much at stake, is there some inertia that keeps us from doing the right thing?

BILL MCKIBBEN: Sure. The fossil fuel industry is by far the most profitable industry on the planet. That provides an enormous incentive to keep with where we are going, at least for the small group of people who benefit immensely from it. Of course in our political systems it doesn't take much of that money spread around to delay and forestall action. For twenty years in the United States, we have had a perfect bipartisan record of accomplishing nothing on climate change. We are not behaving logically. For the moment, our big brain isn't working very well, at least on a kind-of-species scale.

I think the answer is since we can't compete in money with these guys, we have got to find other currencies to work in. Those include building movements, big movements of people that will demand political change. Of course they also include making what changes we can on a local level, building institutions and structures for the world that will come after fossil fuels. We have to act on the personal level. We need to act on the community level. We need to act on the national and global level.

JOHN ROBBINS: Is it frustrating to live with the awareness that you carry, and yet to see the complete lack of a meaningful response to a predicament that is increasingly dire?

BILL MCKIBBEN: Yes, I think at this point it is pretty clear that if left to its own devices, the system won't produce change, so we can't leave it to its own devices. We've got to try and push it. I founded 350.org with seven college students in 2008. We had no money. It's become by far the biggest grassroots global climate campaign. We operate now in every country on earth except North Korea. We are beginning to have some successes. We had a huge campaign in fighting against the Keystone Pipeline from the tar sands of Alberta. Big oil has done its best to push it through. But in 2011 the President denied the permit, and that is because 1,253 people went to jail during what was the biggest civil disobedience action in thirty years on any issue in this country. It is because 800,000 people in a single day sent messages to the Senate. It is because people showed up to ring the White House five deep in people. It is because we mobilized a movement. We are going to have to do that over and over and over again.

JOHN ROBBINS: A lot is said about the role of young people in social change movements, but I think elders also have an important role. It's older people who have the perspective of time, who know that generations come and go, and who know that there are long-term consequences.

BILL MCKIBBEN: Sure, and right now if you are twenty-one, getting an arrest record is probably not the best thing in the world. One of the few advantages of growing older is after a certain point, what the hell are they going to do to you?

JOHN ROBBINS: I think that the situation is so confusing for people sometimes because of the way the oil companies fund denial, fund confusion, and fund this ridiculous refusal to confront reality. Do you get sick of it?

BILL MCKIBBEN: Yeah, I do. But I am very heartened to watch people begin to come together all around the world. Some of our strongest movements around climate change are in places that have done nothing to cause the problem. You can take countries like Bangladesh, where you can hardly even measure how much carbon the country emits. Yet even in Bangladesh, there are people who are willing to take a real role on these issues. I always figure if they can do it, I best get to it also.

JOHN ROBBINS: What about the role of agriculture and food production? Eating more locally grown food reduces the carbon footprint of our diets, and so does eating more organic foods because synthetic fertilizers and pesticides are almost entirely made from oil and natural gas. Getting more of our protein from plants is another great step.

BILL MCKIBBEN: You are exactly right. Eating lower on the food chain helps.

JOHN ROBBINS: Remarkably, all of these steps to a more climate-friendly diet are also steps to a healthier diet.

BILL MCKIBBEN: Well remarkably or maybe not remarkably. It is my experience that things that are good in one dimension are often good in many, and things that are bad in one dimension are often damaging in many.

JOHN ROBBINS: The UN Food and Agriculture Organization (FAO) report *Livestock's Long Shadow* found that 18 percent of the entire world's greenhouse gas emissions stem from livestock production. That is more than the amount attributable to all the cars, trucks, airplanes, and other modes of transportation on earth.

Other research has found the figure to be even higher. World Bank scientists writing in the journal *World Watch* put the figure above 50 percent. Their analysis was based on the reality that about a quarter of the world's land-mass is used for grazing livestock, and about a third of all arable land is used to grow animal feed. They believe that the reforestation of land currently being used in live-stock production would be the fastest and most economical way to dramatically reduce the amount of carbon in the atmosphere.

BILL MCKIBBEN: Food is a huge part of the picture, as one would expect since food is a big part of our lives. Of course cutting down forests in order to get more pasture-land is a major factor. So is the methane belched from the animals themselves, and the energy it takes to ship perish-able food around the world, and on and on.

JOHN ROBBINS: Cattle produce a tremendous amount of methane. The process as you know is technically called eructation, although people call it belching. Belching may sound humorous, but it is not a laughing matter. It is one of the largest drivers of climate change.

BILL MCKIBBEN: In recent years some scientists have begun to ask a different set of questions. There were more hoofed mammals on this continent 300 years ago than there are now, if you add up all the bison and antelope and things. They were busy belching too. The question is why weren't they causing the same kind of problem? The early evidence seems to be that when you have animals eating grass and moving around all the time the way that wolves used to very effectively do, their hoof action and the con-stant deposition of manure creates extremely healthy soils. Farmers are beginning to learn on a very local, small-scale level, which would be the only way to do it, how to mimic some of that.

There is a sort of growing movement of carbon farmers. These are people trying to improve the soil, and showing that you can grow topsoil much more quickly than we used to think was possible. There are lots and lots of possibilities here, most of which boil down to getting agriculture back on a reasonable scale.

We spent the past 150 years moving people off the farm. At the moment in America there are twice as many prisoners as there are farmers. The result is that we have replaced all that human knowledge and labor with oil, with cheap energy, and we've got to reverse that trend. We are never going to go back to 50 percent farmers, but we should be moving a little bit in that direction. I've got to tell you, the best new statistic I have heard in a very long time came from the USDA. Last year they reported that for the first time in about a century and a half, there were more farms in America instead of fewer. That is extremely good news. It points us in a new direction.

JOHN ROBBINS: May that direction grow.

BILL MCKIBBEN: Amen!

JOHN ROBBINS: It takes an average of nine to twelve pounds of grain or soybeans to produce one pound of feedlot beef. Do you think there is any possible defense for industrial animal agriculture?

BILL MCKIBBEN: None whatsoever. Things that are terrible in one aspect are often terrible in others. In order to have feedlot agriculture, you have to grow all that corn to feed animals, which is a highly fuel intensive process. It is clearly an inhumane and unpleasant life for the animal. Once you get that beef, unlike grass-fed beef, it has a much different nutritional profile. Feedlot meat is not good for you and it is frankly too cheap, so we eat too much of it—to our detriment. If we are going to eat meat, it makes far more sense to have locally raised animals, which will be more expensive because you have to pay someone to go out and move the animals around and pay attention to them. Then we will learn to eat meat the way that most cultures in the world do, as almost a kind of condiment instead of a great big honking slab.

JOHN ROBBINS: James Lovelock has said that if we gave up eating beef we would have twenty to thirty times more land for food than we have now, but he is also convinced that humanity has not evolved sufficiently to make the changes that would be needed to avert disaster. When you listen to him and others who feel we may have already passed the point of no return, what do you think and how do you feel?

BILL MCKIBBEN: We face some very depressing reports, and Jim Lovelock shouldn't be taken lightly. He has a strong intuitive sense, and was the first proponent of the Gaia Theory to describe how the world works. It makes sense to listen carefully.

JOHN ROBBINS: I see a tendency to retreat into ever narrower and more destructive forms of self-interest. Do we still have the potential of fashioning an effective global response?

BILL MCKIBBEN: There are times when I think that this really is a test to find out whether that big brain of ours was a good adaptation or not. Maybe it depends whether it is connected to a big enough heart.

I don't think we know the outcome of this. I think if you were a betting person you might bet that we are going to lose, because we have been losing so far. But I think it is not a bet that you are allowed to make. You've got to get up in the morning and figure out what you can do to change the odds quickly.

One thing that I think we are beginning to get across is who the radicals in this scenario are. They are not us. What I want is a world a little bit like the one I was born into. That is a pretty conservative notion. The radicals work at oil, coal, and gas companies. They are willing to alter the chemical composition of the atmosphere. That is the most radical thing anybody ever came up with.

JOHN ROBBINS: You are asking people to wake up, to outgrow denial, to confront the consequences of our fossil fuel orgy. What kind of a response are you finding on a global level?

BILL MCKIBBEN: Most of the people that we work with around the world at 350.org are poor, black, brown, Asian, and young, because that is what most of the world is made up of. I had always heard people say, "Oh environmentalism is something for rich white people. If you didn't know where your next meal was coming from you wouldn't be an environmentalist." It just turns out not to be true. This is a global fight and people on whom the future is bearing down hardest are often the most engaged.

JOHN ROBBINS: There are things we can do by ourselves and close to home, but one of them is not going to be solving global warming. If we are going to do that, it is going to require political action, which means organizing. It means being with your neighbors and with people all around the world in order to make a collective difference. That is hard and the political system is not a fun place to hang out, but it is what we are called to do. It is our challenge the way that the Civil Rights Movement was the challenge in our parents' time and in our grandparents' time they had to go off and fight Hitler. That wasn't much fun either. But it's time for us to do what we need to do. It's time to grow up.

BILL MCKIBBEN: There you go.

9

Ronnie Cummins
Truth about Organic Foods

Ronnie Cummins is founder and director of the Organic Consumers Association (OCA). OCA is a nonprofit network of 850,000 consumers dedicated to safeguarding organic standards and promoting a healthy, just, and sustainable system of agriculture and commerce. Ronnie is one of the top organic food experts in the world, and his most recent book is Genetically Engineered Food: A Self-Defense Guide for Consumers.

In the United States, the market for organic foods grew from $8 billion to $31 billion in the first decade of this millennium. Are they an overpriced scam, or fundamental to your health and the health of our world? And what is the massive difference between "organic" and "natural"?

JOHN ROBBINS: You have significant expertise in understanding the environmental impact of industrial agriculture. Can you speak about the connection between the food on our plates and the world we create?

RONNIE CUMMINS: I think the most serious issue humans have ever faced in our 100,000 to 200,000 years of evolution is the disruption of our climate. We simply have too much carbon dioxide, methane, nitrous oxide, and black soot in our atmosphere. If you look back at the geological history of the earth, there used to be two to three times as much carbon in the soil as there is right now. And where is that carbon matter now that used to be in our soils, that used to be in our forests, that used to be in our grasslands? Well, a lot of it is in our atmosphere.

JOHN ROBBINS: How did this happen?

RONNIE CUMMINS: First of all, we whacked down a large portion of the world's forest. The major cause of deforestation nowadays is not timber companies; the major cause of deforestation is giant agribusiness corporations that clear-cut forests to plant palm oil plantations and to plant genetically engineered soybeans. We have giant cattle ranches that are whacking down the tropical forests in Brazil and elsewhere to run cattle. We've got to stop the deforestation—in fact we've got to reforest the world.

The second thing is that we've got to get away from this gigantic, overproduction and overconsumption of meat and animal products. The grasslands of the United States, where the buffalo used to roam, contained perennial grasses that had roots going fourteen feet or more down into the ground. These held a tremendous amount of organic matter, and the carbon that's now in our atmosphere. We simply cannot continue consuming meat and animal products at the current rate. And those animals need to be raised in the traditional, organic way, eating grass for their whole lives, and in far smaller numbers.

The third point is that we have 385 million acres of cultivated cropland in the United States right now. The way that agriculture was carried out for 10,000 years was an organic-type of cultivation. Starting about a hundred years ago, and really accelerating about sixty years ago, chemical companies realized that if you put toxic chemicals and toxic fertilizer on the soil, you could get a short-term boost in production. So what we have done by pouring billions of pounds of pesticides and chemical fertilizers on the soil is we've killed the living earth. We've destroyed the soil food web below the surface of the earth that used to retain all this carbon organic matter. It's been released up into the atmosphere.

In addition, nitrous oxide, which is 200 to 300 times more damaging to the climate than carbon dioxide, is released from using this chemical fertilizer. Methane, which is twenty to fifty times more damaging to the atmosphere and climate stability than carbon dioxide, is released from these giant animal farms, and from the overproduction of animals for meat. These animals are belching and farting methane at egregious rates.

Not only is the climate becoming more and more chaotic, but we're literally killing the oceans because the ocean is trying to absorb all this excess greenhouse gas from the atmosphere, but it has reached capacity. The ocean is now getting more acidic, and the plankton, which represents the fundamental base of the food chain in the oceans, is dying off at an alarming rate. So we're heading for disaster, not only for humans but also for every living organism, if we don't turn things around. It's not just the coal plants, it's not just the fossil fuel burning cars, it's not just our overuse of things like air conditioning. It is what we eat every day and how we farm. We've got to generate large-scale reforestation, and to get back to the traditional methods of organic agriculture and animal husbandry if we're going to raise animals, or we are going to be a species that is going to vanish from the earth.

JOHN ROBBINS: As the oceans absorb the carbon, they are becoming warmer and more acidic. The phytoplankton are starting to show signs of vulnerability. We could reach a tipping point where there's a sudden die-off of phytoplankton in the oceans. Given the fact that these tiny creatures produce 50 percent of the world's oxygen, the consequences to all oxygen-dependent life on earth could be catastrophic. The immensity of what is actually at stake here is almost mind-boggling. What is keeping us now from making the needed shifts?

RONNIE CUMMINS: One thing we need to look at is that 1 percent of the world's population is responsible for 50 percent of the greenhouse gas pollution. So part of the problem is that the affluent people of the earth have not acknowledged their primary responsibility in causing the problem and in solving the problem. Once we can agree to that, then we can look at the major obstacles.

JOHN ROBBINS: Why are we still using chemical agriculture, and how can we promote a change?

RONNIE CUMMINS: The U.S. Department of Agriculture is supposed to represent the interests of the people, as well as the farmland, pastures, and animals in the country. It has a budget of $100 billion a year. That's Americans' tax money. But almost nothing about it is geared toward the transformation away from chemical and industrial agriculture and factory farms. We need to get control of our destiny and we need to stop subsidizing the destruction of the earth. Instead we need to subsidize the salvation of the earth.

Take a look at the Senate Committee on Agriculture, the House of Representatives Committee on Agriculture, and the people who run our fifty state legislatures. Look at who they're taking money from, and then look at the kind of decisions they're making. They are not screaming at the top of their lungs about the public health crisis, the climate crisis, or the environmental crisis. They're listening to the people who give them the campaign donations.

It's not just enough to walk our talk, to educate our friends, to do what's right in our everyday lives. We've also got to get political and start passing laws that really make a difference.

JOHN ROBBINS: Monsanto and its allies are telling us that only through genetically engineering seeds and animals can we feed the world. The meat industry is telling us that in order to feed our growing numbers we have to raise chickens in the dark with their breasts so heavy they can't walk, and we've got to cram hogs into cages barely larger than the size of their bodies. All of these interests seem to depend on us believing that organic agriculture, food that is produced in harmony with the natural world, can't really feed the population. Is organic agriculture capable of supplying the bountiful harvests we need?

RONNIE CUMMINS: Yes. Hundreds of studies that have been reviewed by the United Nations Food and Agriculture Organization and other scientific research bodies have shown that where the majority of the people live, in the developing world, organic farming methods are actually two to ten times more productive than chemical agriculture plantations. Now in the industrialized world, where a minority of us live, long-term studies by the Rodale Institute and others have shown that organic agriculture basically can produce at the same level as chemical intensive, genetically engineered agriculture. So in the industrialized world we'll be able to produce as much without the collateral damage, and in the developing world we'll be able to produce two to ten times as much. Not only can organic agriculture feed the world, it's the only way we can feed the world.

JOHN ROBBINS: Why?

RONNIE CUMMINS: As a result of climate change, we're seeing more drought and also more flooding. When you have adverse weather conditions, organic agriculture vastly outproduces chemical agriculture. This is because with organic agriculture the soil is alive—it's like a sponge. The soil retains up to 40 percent more water, so organically grown crops are better able to withstand drought. And when you have torrential downpours it doesn't wash away the fertile topsoil. The rain is soaked up in the soil. So given the unfortunate fact that it looks like we're going to go through hundreds of years of disrupted and unpredictable weather, we're going to have to utilize organic agriculture.

A second point is that industrialized agriculture depends on fossil fuel-derived chemical fertilizer and fossil fuel-intensive systems of irrigation, long-distance transportation, and cooling. We're not going to be able to afford to use fossil fuels the way we have been in our food and farming system. So all these point to a future in which most of us are going to be farmers or gardeners like we used to be before the turn of the 20th century. We're going to be eating local, healthy, nutritious, mainly plant-based food that's produced within a hundred-mile radius of where we live.

JOHN ROBBINS: Well that will probably make us a healthier, and leaner, and brighter people. And it may be the only way that is sustainable. Today, though, more than half of the sewage that's produced in the United States ends up being treated and applied to gardens and farmland. Called sewage sludge, it's sometimes presented as a safe fertilizer, something like barnyard manure. It's sometimes euphemistically called biosolids. But as you know, it contains many industrial chemicals, it contains medical waste, and it also contains bacteria and viruses that are resistant to antibiotics. Why is its use on the rise and what will happen if we continue this practice?

RONNIE CUMMINS: Human waste can become non-toxic and a useful fertilizer if it is composted properly. But we're mixing human waste with highly toxic industrial waste, hospital waste, and even radioactive waste. We're contaminating on a long-term basis the soils that we need to grow our food.

If you look over the last 10,000 years, people used to compost human waste and use it again. In China and India they still do to this day. That's not allowed under current organic standards, but you could certainly use properly composted human waste on trees and on pastureland. And we're going to have to do that because we are running out of clean water.

JOHN ROBBINS: Is there a connection between our uses of synthetic nitrogen fertilizers and the pollution of our water supplies?

RONNIE CUMMINS: There is a dead zone in the Gulf of Mexico that has grown to the size of Massachusetts. There are now more than 400 of these dead zones around the world in the oceans. These are a direct result of pouring billions of pounds of chemical fertilizer onto our farmlands and our pasturelands, as well as the runoff from factory farms.

JOHN ROBBINS: Wal-Mart says it's making a huge effort now to carry more organic and locally grown foods. When you hear them say this, what goes through your mind?

RONNIE CUMMINS: Green washing. If you look at the overall ecological footprint of this company, you see that it is based on paying workers, including farm workers, as little as possible so that Wal-Mart can sell "food" as cheaply as possible. And I put "food" in quotation marks because most of what Wal-Mart sells is highly processed, chemically and genetically tainted fare that is little better than eating cardboard.

The reason Wal-Mart can claim to be the biggest seller of organic milk in the United States is because the milk they have been selling is not actually organic. It has been coming from factory farms operated by a company called Aurora that has been slapped on the wrist by the USDA and told, "These factory dairy farms that you're running don't meet the criteria of organic and you shouldn't be calling these products organic." So yes I guess it's better that Wal-Mart has some organic products than none. But Wal-Mart's model is to source food all over the world wherever it's the cheapest, exploiting workers every step of the way, and then transporting it long distances. That's the problem, not the solution.

JOHN ROBBINS: Aurora Dairies was proven to have the deceptive practice of labeling dairy products as organic that really weren't. Horizon Dairies was similarly found to be misleading consumers. These are far and away the largest suppliers of presumably organic dairy products in the country. It raises the issue for a lot of people: Can we trust that food labeled as organic is indeed organic?

RONNIE CUMMINS: Well you can trust most of it as long as we have constant vigilance. But you certainly can't trust food that's coming out of China, even if it says it's organic. And I don't think you can trust anything in a Wal-Mart that says it's organic. We have to realize that the only reason why we still have pretty good organic standards in the United States is because we've fought every step of the way. Back in 1997, the U.S. Department of Agriculture tried to tell us that it would be okay to have genetic engineering and still call it organic, and it would be okay to use toxic sewage sludge and still call it organic, and it would be okay to irradiate our food with nuclear waste or heavy-duty X-ray treatment and still call it organic. But we fought and we won. They've tried to degrade standards on intensive confinement of farm animals, and on use of certain chemicals in food production. We've done a pretty good job at fighting back, but until we have a fully transparent democratic society, we're going to have to keep fighting for these standards.

Most people think that natural is the same as organic, but just cheaper. Unfortunately "natural" on a label basically means nothing.

JOHN ROBBINS: You know that our current agricultural practices are destroying the environment and our lives in countless ways, and you also know that it doesn't have to be this way, that there really is an alternative. You know how much better things could be. How do you live with what you know?

RONNIE CUMMINS: I remember in the late 1960s when a lot of us realized that we had to build examples of the new society that we believed in if we were ever going to convince the majority of the people to go along with it. So we started food buying clubs in cities all over the country. And then others of us moved out to the country and started rural communes. I'm not sure a lot of us realized that forty years later we were going to have a multibillion-dollar powerhouse alternative food and farming culture that's poised to become the norm instead of just the alternative.

Each step of the way, what has inspired me is to see people who've not only correctly identified the problems, but who also are actually living the solutions. In spite of all the enormous problems around the world, I think we are on the track to a world transformation. You don't notice how close we are to this tipping point until you actually get there. But I think we are close. We're already doing everything on a smaller scale that we need to do on a larger scale.

Steps You Can Take:
Food for a Healthy Planet

Grow Food

It takes time, but gardening is the most economical way to enjoy the freshest possible food. In urban neighborhoods, community gardens are a great way to grow food and build community at the same time. There are an estimated 18,000 community gardens in the U.S. and Canada. For resources to help you start one, visit the American Community Gardening Association at *www.communitygarden.org*.

Buy Direct from Farmers

Supporting farmers' markets is a great way to get access to fresh, healthy, local food, and to support local living food economies. It can be fun, too. When you join a CSA (Community Supported Agriculture), you enter into a direct win-win partnership with local farmers. In the U.S., the number of farmers' markets has more than doubled in the last decade. For more info on farmers' market and CSA opportunities near you, visit *www.localharvest.org*.

Use Bulk Bins

When you buy food, you're also buying whatever packaging it comes in. And though you may eat the food, chances are, the packaging will wind up contributing to pollution. Buying beans, whole grains, and other nonperishables from bulk bins will reduce packaging, and it also saves you an average of 56 percent over buying the same items prepackaged.

Chomp Climate Change

Since cows impact our climate more than cars, the most powerful thing an individual can do to take a bite out of climate change is to eat a more plant-strong diet. In fact, a vegetarian who drives a gas-guzzling SUV may have a smaller "carbon footprint" than a meat-eater who drives a Prius.

Resources for Sustainable Food

350.org

www.350.org

Founded by Bill McKibben, 350.org is a global grassroots movement to solve the climate crisis and push for policies that put the world on track to get the atmospheric proportion of carbon dioxide to 350 parts per million. On the 350.org website, the 350 Food and Farm section explains the role of industrialized food production and livestock in the climate crisis. The site introduces ways to get involved with numerous climate change related campaigns and projects, such as: Ending Fossil Fuel Subsidies; Connect the Dots: Showing the Human Face of Climate Change; Climate Change Workshops; and Activist Groups around the world.

Organic Consumers Association

www.organicconsumers.org

Founded by Ronnie Cummins, the Organic Consumers Association (OCA) is an 850,000 member online and grassroots nonprofit public interest organization campaigning for health, justice, and sustainability. The website is a portal for current news events, articles, and OCA-sponsored campaigns and projects. The website offers links to local events, organizations, news, and businesses; the weekly "Organic Bytes" newsletter; an organic buying guide; a Take Action page; and more.

Chomping Climate Change

www.chompingclimatechange.org

According to this site, the common view of climate change requires fossil fuels to be replaced quickly with a renewable energy infrastructure—and that's now estimated to cost $18 trillion and take decades to enact. Here comes a fresh view promoted by Chomping Climate Change!: The widely agreed goal of stopping global climate chaos can be almost fully achieved by replacing 25 percent of today's least eco-friendly food products with better alternatives—quickly and inexpensively.

Bioneers

www.bioneers.org

Bioneers is a hub of breakthrough solutions and leading-edge ideas that celebrate the genius of nature and human creativity to create a healthy planet. The website connects countless innovators such as farmers, educators, scientists, and social justice activists with engaged citizens who are making a real difference to create change within their communities and workplaces.

Meat-Free Monday

www.meatfreemondays.com

Founded by Sir Paul McCartney, and Stella and Mary McCartney, Meat-Free Monday is a global campaign calling on people to give up meat every Monday as a way of "doing your bit to help protect our planet." They offer educational resources, tips, and inspiration to help you participate or spread the word.

Slow Money

www.slowmoney.org

The Slow Money Alliance brings people together to match investors with entrepreneurs who focus on saving farmland. Slow Money Alliance is committed to supporting a new generation of small and midsize organic farmers; rebuilding local and regional food processing and distribution; improving nutrition and otherwise remedying the imbalances of a food system that they consider too consolidated, too global, and too industrial.

Sierra Club's True Cost of Food Campaign

www.sierraclub.org/truecostoffood

The True Cost of Food is a campaign to promote sustainable food choices, hosted by The Sierra Club National Sustainable Consumption Committee. The website promotes the idea that the consumer, through food choices, can stop the practices that harm our health, our planet, and our quality of life.

Chefs Collaborative

www.chefscollaborative.org

Chefs Collaborative is a national chef network focused on changing the sustainable food landscape. The website offers a programs page, seafoods solutions, information on sustainable foods including a link to the annual Sustainable Foods Summit, a page for special events, and a blog.

Beyond Pesticides

www.beyondpesticides.org

Beyond Pesticides works with allies in protecting public health and the environment to lead the transition to a world free of toxic pesticides. Beyond Pesticides provides the public with useful information on pesticides and alternatives to their use. The website provides a daily news blog, educational resources, and information about handling a pesticide emergency.

PART IV

Humane Food for a Compassionate World

A century ago, the typical dairy farm had about twenty cows. Today many dairy farms in the developed world have more than 10,000. The large-scale livestock industry is trying to pass legislation to make it illegal to take photos or videos of the conditions in which their animals are raised and slaughtered. What don't they want us to see?

Most people want food that is humanely produced. So much so that more than 71 percent of the American public supports undercover investigations of livestock operations. Suppliers are slowly beginning to notice and take small steps.

And yet a lot of the time, labels like "humane" or "cage free" are really just a bunch of BS.

What's really going on in the struggle for compassionate food? These experts will help you to get informed, get empowered, and become an advocate for compassion.

10

Gene Baur

Changing Hearts and Minds about Animals and Food

Gene Baur is president and co-founder of Farm Sanctuary, the leading farm animal protection organization in the United States. Hailed as "the conscience of the food movement" by Time *magazine, Gene has been raising awareness about the abuses of industrialized factory farming for more than twenty-five years. He has a master's degree in agricultural economics from Cornell University, and has personally conducted hundreds of visits to farms, stockyards, and slaughterhouses, researching and documenting conditions. Gene's 2008 book,* Farm Sanctuary, *is an international bestseller.*

You may have seen some videos on how factory farms treat livestock. You may have seen how unclean, unsafe, and inhumane the conditions are. But do you know how they affect your health and life? Do animals really have feelings, personalities, and even the ability to heal from trauma? What happens when abused animals are rehabilitated? Are "free-range" or "natural" animal products really better? If you want your food choices to support a more humane world, what can you do?

JOHN ROBBINS: You are a pioneer in the field of undercover investigations. As you've uncovered what's really going on in farming operations, what have you found out, and why are you so passionate to expose the truth?

GENE BAUR: Most people would be appalled by the conditions that are commonplace on today's factory farms. For this industry to continue, it must hide its cruelties from consumers who are unwittingly supporting it. In fact, the industry is now trying to pass legislation that would make it illegal to take or distribute pictures or videotape of the conditions at these facilities. That says an awful lot about just how bad the conditions are, and the fact that they are unacceptable in our society.

On factory farms, animals are treated like production units—not like living, feeling creatures. It is common for them to be locked in warehouses by the thousands. They are packed into cages and crates so tightly they can't even turn around or stretch their limbs. This is an affront to the animals and to our humanity. Farm animals should be free to go outside, to engage in normal behaviors, and to develop social bonds and relationships with other animals.

We believe that most people are humane. Most people oppose cruelty. Most people do not think it is right to treat these animals with such wanton disregard. A big part of our work is educating and raising awareness about the realities of industrial animal farming and animal slaughter, and encouraging people to make food choices that align with their own humane values. So instead of saying, "Don't tell me, I don't want to know," people should be able to look and know exactly where their food comes from and feel good about it. We encourage people to make choices that are aligned with their own values and also aligned with their own interests.

Modern meat production is creating horrible suffering and tragic deaths. It's causing the deaths of billions of animals, of course, and it is also causing people to suffer and die from preventable illnesses. We think that by making informed choices, people can make a world of difference for themselves and for other animals.

JOHN ROBBINS: We have laws against cruelty to animals in the United States, but in most states, the legislation specifically exempts animals destined for human consumption. The result of this is that the animal agriculture industry routinely does things to animals that, if you did them to a dog or a cat, would get you put in jail. What can we do to help put an end to the most abominable practices of modern meat production?

GENE BAUR: Most of the anti-cruelty laws exempt farm animals as long as the practices are considered to be normal by the agriculture industry. Bad has become normal, and no matter how cruel it is, normal has become legal. If you want to take action, the first thing you can do is to stop supporting those industries with your money. Don't buy animal products, especially those that come from factory-farmed animals. Ideally, you can move towards a vegan lifestyle. But begin by saying no to products from factory farms, which comprise roughly 99 percent of the animal foods sold in the United States today.

It is also good to get to know your elected officials and to educate them about your interests and concerns. Enacting legislation is a long and tedious process, and laws that are passed are rarely as strong as we would like. But it is still important to get to know your elected officials and to engage in the process. Express your concerns and your desire for there to be some oversight, transparency, account-ability, and basic standards on farms. Ask lawmakers to support legislation to require, for example, that animals be given the right to at least turn around, stretch their limbs, and engage in basic natural behaviors.

JOHN ROBBINS: Gene, you and I have been involved in this for more than a quarter century. What has surprised you in your years of activism?

GENE BAUR: I am sometimes surprised by how habits can be so deeply held and so slow to change. People often do things because it's what they grew up doing, it's what they've always done, and it's what people around them are doing. I am sometimes surprised by how afraid people can be of trying something new. Too often, change comes only on the heels of a disaster like a heart attack or some other life-threatening illness. I find it inspiring and refreshing when people don't need to experience a drastic incident to step back, assess their way of living, and decide to make choices that make sense. When people want to, they can change in an instant.

I am encouraged to see more and more people becoming aware of the problems with animal agriculture, and adjust-ing how they eat. For the first time, the number of animals being slaughtered in the United States has started to go down. There are more vegan restaurants and vegan foods available than ever before. I think consumers are starting to come around.

JOHN ROBBINS: Some people have the fear that without a lot of meat in their diet, they won't be healthy and strong. Yet you just ran a plant-fueled first marathon at nearly the age of 50, and Ocean and I have run veggie-powered marathons together starting when he was 10 years old. People who eat a plant-strong diet are known to have lower rates of heart disease, obesity, diabetes, and many forms of cancer—and often to have more energy and vitality, as well. And yet still the entrenched thinking in the larger culture suggests that vegetarians are taking a risk by not eating meat.

GENE BAUR: When everybody around us is doing things a certain way, we are influenced by that. We are social animals, and we tend to do what those around us do. I grew up eating meat, largely because everybody around me was eating meat. I developed the meat eating habit without really thinking about it, but as I learned about our food system, I decided to go vegan. A growing number of people are starting to live and eat in a way that is more conscientious, and that has a ripple effect. We rub off on those around us. I think we are in the midst of a grassroots food revolution underway right now, and it's very exciting.

JOHN ROBBINS: I have heard many people tell me that they have been deeply moved by visiting one of Farm Sanctuary's shelters. You have a very large farm in Upstate New York, an even larger one in Northern California, and a third near Los Angeles.

GENE BAUR: When people come to the farms they get a chance to interact with cows, pigs, chickens, turkeys, sheep, goats, and ducks. They get to see these animals as living, feeling creatures, and to recognize that they have personalities, just like cats and dogs. They have wants, they have desires, and they have friends. They develop deep relationships with their herd mates. Cows will bond with other cows or sometimes with sheep. Sometimes you have different species connecting with each other.

JOHN ROBBINS: The animals that you have in your sanctuaries have, for the most part, been rescued from stockyards, factory farms, and slaughterhouses. You rehabilitate them and then you provide them with lifelong care. How are you able to help these animals recover from a lifetime of abuse and neglect?

GENE BAUR: It is really an amazing thing to watch animals recover and begin to enjoy life and flourish. Prior to coming to Farm Sanctuary, these animals had known only cruelty and abuse at human hands. When they first come they are often very frightened, because in the past when they have been approached by a person, it has meant pain. But once they come to Farm Sanctuary, they are treated with kindness. They also see the other animals at the farm, who welcome them and let them know it is a safe place. They start healing from their physical and emotional injuries. It is a wonderful thing to see an animal who was so injured or sick that they couldn't even stand start being able to stand and then walk and eventually play and experience joy for the first time. That is what happens at the sanctuary when we provide animals with a healthy physical, social, and emotional environment. The animals are very resilient, just as I believe humans are very resilient. When you start taking care of yourself, amazingly positive things can happen.

JOHN ROBBINS: I agree that animals have emotions and an emotional life. But not everybody in our society holds that thought. What is your reflection on that issue?

GENE BAUR: I think one of the reasons that people tend to believe farm animals don't have feelings and emotions is because of the way that we treat them. Humans have an abusive relationship with farm animals. We don't want to see ourselves as cruel, so when we are acting with cruelty to other animals, we want to rationalize it and say, "Well those animals really don't have feelings so it doesn't really matter." I think this sort of mindset makes us feel more comfortable with the way that we are mistreating other animals, but it also closes us off. Cruelty to animals is bad for the animals, but it is also bad for us. It requires us to cut off some of our empathy.

You see this in other circumstances too. When one group of people is abusing another group of people, there is a tendency for them to denigrate the abused group, and to come up with reasons why certain people don't deserve to be treated well. In the case of farm animals, one of the rationalizations we develop is that they don't really have emotional lives. This rationalization is only possible if we don't actually spend time connecting with the animals. One of the things that happens at Farm Sanctuary is that people get a chance to know these animals and recognize that they do have emotional lives. Farm animals are not that different, in many ways, from our cats and dogs.

JOHN ROBBINS: It is a wonderful thing when people come to feel an emotional connection with animals, because then we feel less alone and less isolated. We are restored to our interconnection with the web of life and other living creatures. We become more available to receive the gift that animals can bring to our spirits and of playing the part that we can play in their lives. The human-animal bond is profound to our lives, and when we turn it into a relationship of exploitation rather than one of respect, I think we do something tragic to ourselves. Teaching a child not to step on a caterpillar may be as important for the child as it is for the caterpillar.

GENE BAUR: I could not agree more. Kindness to animals is good for animals and it also says a lot about us. Farm animals are so much at our mercy. To treat them with the sort of disdain and abuse that we do is not consistent with our humanity, with our better selves, or with our empathy. I think empathy is a natural human characteristic and a very important quality. The factory farming industry and the animal slaughter business require us to shut off our empathy. When we do that, I think we do not live as full a life as we could.

JOHN ROBBINS: The industry is treating animals this way for short-term profit, but they have to pull the wool over people's eyes to get away with it. They spend a lot of money—and, frankly, genius—on propaganda. They give us images of happy cows and Lassie and Timmy running around on the farm. They don't want people to know how it is really done today.

But consumers are becoming increasingly aware and concerned. In response, animal producers and food retailers are developing programs to market their products in ways that appeal to these concerns. We see labels like "Humane," "Natural," "Cage Free," and "Organic." But it is very difficult to know what these terms actually mean. Can you give any guidance to people who want to do the right thing and are befuddled by the labeling confusion?

GENE BAUR: There is a growing aversion to the cruelty of factory farming, so an increasing number of animal producers are starting to label their products as if the animals are being treated well, but unfortunately those labels tend to sound a lot better than they really are. In the case of free-range for example, most people would assume that animals are outdoors living a natural life. But in fact free-range only requires that animals be given access to the outdoors. Access is not defined, so in many instances what it means is that you have a factory farm warehouse with thousands of animals in it, and there is a small door that goes to a crummy little paddock outside. That is technically access to the outdoors, but the animals never use it and they are basically confined in a factory farm. That is an example of what free-range might mean.

In the case of natural, the word means nothing about how the animals are raised. It only describes the processing after the animals are killed. You can have beef cattle, for example, that are raised in a feedlot, given hormones, and fattened for slaughter just like other factory-farmed animals—and then their meat can be sold as natural. These terms unfortunately mislead consumers. The best thing to do is to eat a plant-strong diet, and also, as much as possible, to go to farmers' markets to buy food that is local and organic. It's good to get to know the farmers, and to get connected with the source of your food.

JOHN ROBBINS: You are asking people to take responsibility for their actions and to bring those actions more into alignment with their hearts. People don't always like being asked to do that. What keeps you going in the face of resistance?

GENE BAUR: I am basically an optimist. I believe that most people are humane and want to live on a healthy planet. Most people want to eat wholesome food that makes them feel good instead of food that makes them sick. I have faith that people will ultimately make decisions that make sense. With the Internet, information is more available now than ever, and people are able to make more informed choices. We will just keep on plugging away, day after day, year after year, and bit by bit we are going to keep making progress.

11

Nicolette Hahn Niman
Eating Right and Righteously

Nicolette Hahn Niman is an attorney, a livestock rancher, and the author of Righteous Porkchop: Finding a Life and Good Food Beyond Factory Farms. *A frequent contributor to the* New York Times, *Nicolette has served as senior attorney for the Robert F. Kennedy-founded environmental organization Waterkeeper Alliance. She is an inspired advocate for CSAs, school lunch revamps, school gardens, networks of pasture-based farms, and farm to table restaurants. Nicolette is a vegetarian, and yet she and her husband, Bill Niman, operate a natural, organic meat company called BN Ranch, and Bill also founded one of the first and most prominent grass-fed ranching companies, Niman Ranch. Whether you call yourself a vegan, a carnivore, or a bonsai tree, Nicolette shares vital information that will make a difference in your life.*

How do changes in food production impact your world? And what's the skinny on grass-fed beef? Can it really be more humane and sustainable?

JOHN ROBBINS: In the first part of your fabulous book, *Righteous Porkchop*, you describe working closely with Robert F. Kennedy, Jr. in the effort to expose and reduce the pollution from factory meat production. As you undertook this work, what did you learn about the environmental impact of industrialized animal agriculture?

NICOLETTE HAHN NIMAN: We've shifted from a system where a lot of individual independent farmers were raising their animals in fairly small flocks and herds. The animals generally used to live outdoors and on grass. In some climates they would spend a lot of the winter in the barns, but they would still be on deep-bedded straw, and they would go out on sunny days and be out all the time in the nicer parts of the year. That system was once the dominant method of raising animals for food.

In the mid 20th century, the United States shifted from that more traditional sort of pastoral system to a much more industrial system. The animals began to be treated very differently, and this had significant environmental consequences. The animals were essentially brought indoors, brought in off the grass, and kept in very large and very crowded herds and flocks.

In the beginning of the 20th century, a typical dairy farm had about twenty cows. Today many dairy farms in the United States have more than 10,000. At the beginning of the 20th century there were a few dairy cows scattered all over the landscape at pretty much every farm. Now it's just a small number of farms, but they have these huge herds that are living in buildings. They're spending most of their lives standing on concrete. Because of the concentrated population and because they're kept continually indoors, the big environmental issue is their manure. In dairy operations and many pig operations, that manure is actually liquefied. It's made into a liquid form so that it can be moved around more easily. It's often stored in huge, multimillion-gallon ponds. That is just inherently environmentally hazardous, because the liquid manure gets into the air, it gets into groundwater, and it spills into rivers and streams.

What I was doing as an environmental lawyer for Bobby Kennedy was to try to address that environmental problem.

JOHN ROBBINS: Did you have success?

NICOLETTE HAHN NIMAN: Initially we were really focused on litigation, and we launched several lawsuits that lasted over multiple years. We had some very good decisions along the way and they resulted in some important remediation. But I think our greater success was that we began a more nationalized connection of grassroots activists. We found pockets of people in different places all around the country trying to fight factory farms in their local communities. We began to network them, and to play a part in what has become an important national movement to try to de-industrialize the way our food is produced.

JOHN ROBBINS: It seems that many of our agricultural policies subsidize factory farms and massive feedlots, but don't support smaller scale, more humane, more sustainable approaches. Would you say that the playing field has actually been tilted in favor of industrialized meat production?

NICOLETTE HAHN NIMAN: I think that's true, and that historical context is important in understanding how we got to this point. It didn't happen overnight and it didn't happen on its own.

Government policies were put into place shortly after World War II that were designed to increase production. There were some pretty significant food shortages especially in Europe at that time, so it made sense. Our policies focused on increasing production to the maximum degree, and that's why the factory farm was deliberately subsidized by the government. But after a decade or so, there weren't significant food shortages any more. We actually had over-production problems of most agricultural products in the second half of the 20th century.

We did a lot of things to increase productivity and to bring down costs, and that was in one way sort of laudable. But in other ways it caused tremendous collateral damage.

We created what economists call the externalization of costs. We created agriculture that was extremely polluting. We no longer preserved topsoil, which is probably the single most important natural resource when we're talking about food production. We began to create an agriculture that was strongly dependent on antibiotics and other drugs. And we created conditions for animals that the vast majority of Americans consider inhumane when they actually see them.

JOHN ROBBINS: Was there an event in your life that propelled you into being as passionate as you are about this?

NICOLETTE HAHN NIMAN: I was hired by Bobby Kennedy to be the senior attorney for the Waterkeeper Alliance in New York. After I arrived at my job, Bobby approached me with the idea of working full-time to launch a national campaign on this issue. I didn't think it sounded very appealing because I kind of thought, "Oh great, I get to work full-time on poop." I could see myself staying up late all night reading manuals about methane digesters. I expressed some hesitation about it and Bobby said, "Well what you need to do is go visit some of the communities where these big factory farms are being placed and meet the people, and see the conditions."

I went to Missouri and I met people, many whose families had lived in the community for multiple generations. They were small-scale farmers and they were showing me what had happened to their community with the advent of the big livestock operations. I heard how the waters had become contaminated, and about the way they could no longer sit on their porches because the odor was so repulsive. They couldn't hang their laundry outside anymore, because it would get the smell of manure on it. It was a shocking experience to talk with people who could no longer enjoy their own homes and their own farms. They were raising livestock themselves, but they were describing a method that they found incredibly inhumane and offensive, as well as causing pollution and odor.

I had actually been a city commissioner in Kalamazoo for four years before this time, and I was very clear on the role that local governments had in controlling land use. But in Unionville, Missouri, they told me they had actually passed local ordinances to protect themselves from big hog operations. The Premium Standard Farms company wanted to locate in their community, so the company simply went to the state capital and lobbied for an exemption from the local ordinance—and they were given it. So the operation just came right in, in spite of the fact that the community was very unified in not wanting it, and began building these big hog factory farms.

When I visited, there were about 75,000 hogs within a two-mile area. This was a community that had done everything it could to protect itself, but they couldn't do anything against this incredibly powerful agribusiness corporation that just steamrolled over them.

To me it was shocking on a lot of levels. It was shocking as an environmentalist. It was shocking as someone who believed in every American citizen's right to enjoy their own property. And it was shocking, and haunting, to see the way the animals lived.

Seeing the whole picture unified all these things that I always cared about. I had always been interested in animals. I had always been interested in the environment. And I cared a lot about local control and self-governance. When I saw what was happening, I realized this issue was something that was going to become a life's passion.

JOHN ROBBINS: What can individuals do to help?

NICOLETTE HAHN NIMAN: People often have a sense that this problem is so big that they can't really do anything about it. I get this a lot from people within the agricultural community, who are bought into the more industrial model and then say, "You know I actually really agree with what you're saying, but I don't see any other way."

I'm trying to help illuminate the path to a better form of food production and particularly a better way of raising animals. I think that everybody involved can have a huge role to play in transitioning towards more sustainable methods. Unlike most other social issues, food is something that all of us are very directly connected to.

There are two important ways that we can affect this. First of all, we are all citizens, so we owe it to ourselves and to future generations to be demanding that our government is actually encouraging good farming practices and food production methods that produce healthy, safe food that can we feel comfortable eating and feeding to our children.

The public policy side of this is incredibly important. For example, there has been a law introduced in Congress every year for about the past ten years that would limit the overuse of antibiotics in animal agriculture. It doesn't say, "You can't use them," but it would reduce the usage significantly, because you couldn't continually add it to animal feed. This has been the law in the European Union for a long time, and every public health official knows that it should be the law in the United States, as well.

JOHN ROBBINS: The routine use of antibiotics breeds antibiotic resistant bacteria—which is now killing more than 40,000 people every year in North America.

NICOLETTE HAHN NIMAN: Yes. But the pharmaceutical companies and big agribusinesses have stopped this legislation from getting passed, every single year. Every American citizen should be telling their representative and their senator that they want them to support this law. It's a perfect example of the kind of thing that we can do as citizens.

Now on the other side of the coin we're all consumers and again, this is really unique to the food issue. We have a tremendous amount of power. You might buy a pair of running shoes once every sixteen months or something, but you probably consume food every day. You might buy it more than a couple of times each week. Every time you buy food you are voting for one food system or another. While it's difficult to completely change your habits overnight, I think all of us can begin examining what we are eating, how it was produced, and how we can support the kind of agriculture and food system that we want.

JOHN ROBBINS: Some of our readers are vegetarian and others want to find humanely raised and earth-friendly animal products. I know that you and your husband, Bill, now run the BN Ranch. The company Bill founded, Niman Ranch, still bears his and your last name. But since he founded it, Niman Ranch has changed. Can you talk about that?

NICOLETTE HAHN NIMAN: Bill had a very stringent standard that he always insisted on that the husbandry lived up to a certain standard for every animal throughout its life. At the time that Bill decided to leave Niman Ranch, the beef protocols were being changed and that was really a major sticking point for him. So he is no longer comfortable with the way beef cattle are being purchased for that company. But as far as I know, the pork and the lamb is all being raised the same way, and honestly I think Niman Ranch and the other natural beef companies are doing a better job than the vast majority of the beef companies in the United States today—but they're just not doing quite as well as we want them to.

JOHN ROBBINS: So you have a high bar in terms of animal treatment, feeding standards, and so forth?

NICOLETTE HAHN NIMAN: We actually view it as part of our life's work. Bill of course founded the company, Niman Ranch, but he was also a member for two and a half years on the Pew Commission, the national commission that looked at animal food production. They made recommendations to Congress and other bodies. As a part of that process he toured many, many factory farms. He'd already been in lots of animal operations over the years. But it became clear to him that it was an important part of his life's work to help show how animals really should be raised. Both he and I have a very strong commitment to the highest possible standards for animal handling, but also for food safety and food quality.

Our feeling is that meat and other foods from animals are something that you don't have to be consuming nearly as often as most Americans are. But if you are going to consume it, it's something you should make sure is produced in the best possible way, because there are so many negative implications if it's not done that way. Shorthand we always just say: Eat less meat, eat better meat. We think that if people just began to shift their diet in that direction, it would be a huge plus for human health, for the environment, and for animals.

JOHN ROBBINS: It's hard to be commercially successful raising farm animals without resorting to cruel practices. At present it's kind of a niche market, and we have so many policies that favor the large-scale factory farm that it seems like you're swimming upstream. What needs to happen so that the playing field can be tilted in the right direction?

NICOLETTE HAHN NIMAN: It is very difficult to do things this way in the current economic context that favors industrial production. That's why if you meet people who are raising animals on pasture, they tend to be maverick types. It takes an independent spirit and someone who's very persistent and determined to do things differently than the vast majority of what's being done out there.

People are eating an average of more than 160 pounds of meat per year in the United States. That's far more than people should be eating. I think our goal doesn't have to be to try to maintain the current levels of production. But I think we should be trying to produce food that is healthful to eat, in a way that is environmentally sound, treats animals respectfully, and provides a good living for people doing the work. I actually think that the Niman Ranch is a good model for that. If public policy were more helpful to them, and didn't provide all the subsidies to the more industrial model, I think they'd probably be enjoying even more success.

The main thing that the Niman Ranch model has shown is that farmers can join together and consumers can seek out that kind of product and say to themselves, "Well, I won't eat as much of the pork or whatever, but I'll eat the best. I'll eat the stuff that I know I can feel good about."

When consumers make that commitment, they are supporting a healthier kind of farming, which I'm hopeful can become the dominant model. Part of the reason I'm optimistic is that the industrial model is truly unsustainable. I mean it is propped up not just by public subsidies, but also by the use of antibiotics, which the public health community is saying we absolutely cannot continue. That will have to stop and similarly the entire food system right now is absolutely dependent on cheap fossil fuels, which we know are not going to continue forever.

The kind of agriculture that we're advocating for is not dependent on massive inputs of fossil fuels or cheap water or agricultural chemicals. It's a system that we have to move towards, given the natural resource limitations that are on the horizon. I believe that this shift will occur. The question is: how are we going to get there, and how soon are we going to get there?

JOHN ROBBINS: You yourself are a vegetarian. What's it like for you to be involved in animal agriculture as a vegetarian?

NICOLETTE HAHN NIMAN: Really good sustainable farms very often have animals on them and they play a critical role. So I didn't have a fundamental objection to what Bill was doing, but I still didn't imagine myself ever being married to someone involved in it, or eventually getting involved in it myself. It was a gradual process where the more I learned and the more I saw, the more comfortable I became. When I moved out to California after we got married I still thought, "I will be an environmental lawyer living on this ranch, but I will not be a rancher." Then what happened was that as I spent more time on the ranch, and as I lived alongside our animals and saw how the land was stewarded, I became not just comfortable with it—I actually became very proud of what we were doing. I wanted to become a part of it.

Bill never tried to draw me into it. It was really my own decision, but it's been a huge pleasure for me to work here, because I feel so good about what we're doing and I think we're playing a very important role as an example of the way things can be done.

JOHN ROBBINS: On a personal level, you are a vegetarian who has married a sustainable ranching icon. Many people I know have difficulty having constructive communication and positive relationships with anyone in their family who doesn't eat the same way they do. How have you overcome these challenges?

NICOLETTE HAHN NIMAN: There was an article on Valentine's Day a couple of years ago for the *San Francisco Chronicle* about unusual couples and we were one of them. We were featured because our story does seem surprising. But one of the things that we emphasized when we were interviewed for the *Chronicle* was that so much of it for us is just about mutual respect. Bill has never tried to turn me into a meat eater and I have never tried to turn him into a vegetarian. At the same time, I have had a growing appreciation for what he's done, and he has had a growing appreciation for the emphasis I've placed on very limited consumption of animal products and the idea of the highest possible care being given to every individual animal.

I think Bill's diet has changed pretty dramatically. He eats a lot less meat than he did when I first met him, and both of us have evolved our eating into a more thoughtful and healthful direction. It's a mutual evolution and I think the foundation of it is in respecting the other and understanding what our common values are. Food is actually the foundation of a lot of the joy we experience on a daily basis.

We're big believers in sitting around a table together. Now we have a young son and we try to eat at least two meals a day together around a shared table. We believe in food as one of the most important ways we commune with one another every day. The common ground is much greater than the differences of opinion.

12

Rory Freedman

How to Stop Eating Misery and Start Looking Fabulous

Rory Freedman is the author of Beg: A Radical New Way of Regarding Animals *and co-author of the three-million copy bestseller* Skinny Bitch: A no-nonsense, tough-love guide for savvy girls who want to stop eating crap and start looking fabulous. Skinny Bitch *spent three years on the* New York Times *bestseller list, and has been translated into twenty-seven languages. Rory has appeared on dozens of national and international television and radio programs, and was named* VegNews *magazine's first ever Person of the Year. She has a deep sense of ethics and conviction that underlie her life and work.*

Find out Rory's keys for reaching new audiences with messages of compassion and health.

JOHN ROBBINS: I've heard you say that when you were a little girl, you would rearrange the stuffed animals on your bed to make sure none of them had their feelings hurt. That struck me, because I did exactly the same thing when I was a little boy. Why do you think most people lose touch with the innocence and the compassion they felt as children?

RORY FREEDMAN: I think one of the first reasons is that parents just don't know any better and they think they have to feed their children meat. Among the first words that children learn and are excited about are animal words. Then when they're being fed animals and they make that connection, parents often lie and say, "No, no, no. It's different chicken, it's not the same." I think it starts as early as that, and then later in life we become addicted to those foods and we start lying to ourselves and don't ask ourselves, "What is this food that I'm eating?" We don't acknowledge that the food on our plate used to be a living, feeling being.

JOHN ROBBINS: So many people who love animals have pets that they consider part of their family. They give their pets names, buy them food, pay their vet bills, and they may even sleep in the same bed with them. And yet they may continue to eat meat from other animals whose lives have been nightmares of suffering. Where does this disconnect come from?

RORY FREEDMAN: I'm not sure I even know myself. I know that I had that same disconnect. I was raised by parents who loved animals and we had pets, and every night we sat down to dinner and ate meat. My parents were kind of health conscious, too. So, somehow that never got on their radar and it certainly wasn't on my radar. In college I met someone who is now one of my oldest and dearest friends. We were trading stories, trying to get to know each other. We bonded over how much we loved animals. Eventually she said, "I'm a vegetarian," and my response was, "Oh my God! I love animals. I'm the biggest animal lover, but I could never be a vegetarian."

I really had no idea what was happening to these animals. I think that's why work like yours and the work that so many animal rights organizations are doing is so important. Because we live in the disconnect, but most of us are compassionate and once we do see exactly what it is we are contributing to every time we sit down to eat meat, we really have a change of heart. Suddenly the addiction is trumped by the compassion and the connection.

JOHN ROBBINS: What changed for you?

RORY FREEDMAN: I had a huge "a-ha" moment when I was in college. One day I got a magazine in the mail from People for the Ethical Treatment of Animals. In this magazine was an article about factory farming and slaughterhouses, and there were pictures.

I had never in my life even considered for a second what was happening to cows, chickens, and pigs, so that I could eat them. As I looked at those pictures, I felt such sadness, horror, and shame that I was part of that. I just knew that I couldn't be a part of it ever again.

I had meat in my freezer, and had planned on eating meat for dinner that night. But that was it. That was the end of that era of my life. From that minute on, I never again ate another cow, chicken, or pig.

JOHN ROBBINS: In your book you describe how the animals are treated in factory farms and in slaughterhouses. I've done the same in my books, and find that most people, if the veil is lifted and they actually see what is going on, are appalled. You don't have to be an animal rights activist or a vegetarian to want your life to be a statement of compassion and to want your life and your way of eating to be in alignment with your caring. Yet, so many people don't want to look—they don't want the veil to be lifted.

If you saw hidden camera footage of a waiter peeing into the coffee pot, you wouldn't say it's fine to drink the coffee as long as you didn't know that he peed into it. But there are people who eat factory-farmed meat who say they don't want to hear descriptions, they don't want to see what's really going on. As though it is okay to eat it as long as you don't have to look at what is happening.

RORY FREEDMAN: One thing I say to people is, "You don't want to see it, but you'll eat it?" I think it really is just the most convenient way for people to avoid making changes—human beings usually do not like change. But if people had any idea of what was really going on, then they would be right on board. Yes, they would be bummed that they had to stop eating certain foods that they liked, but they would find other foods that taste just as good, if not better. I think that's the thing that most people don't realize. It sounds and seems like it is going to be this huge sacrifice. But once you actually do it, it's like, "Oh my God, this is the best thing I've ever done. Why did I think this would be hard? This is fantastic."

JOHN ROBBINS: Do some people think that the diet you recommend is all about deprivation?

RORY FREEDMAN: Oh my gosh, people think that all the time. I just laugh and say, "You will never meet a more food-obsessed group of humans than vegans. Nobody talks about food, likes food, or spends more time eating food, comparing food, dreaming about food, and planning food. We love to eat everything from junk food to gourmet food to healthy food to seasonal food to organic food to sometimes not-so-healthy food and everything in between. Is that deprivation?"

Any change is going to take a little bit of getting used to. I always tell people, "Just give it one month, just give it a try.

Just commit to trying it for thirty days and see how you like it. Every single person who does this says, "Oh my god, I am never going back. I can't believe how good I feel, how much lighter I feel, how much more energy I have, and how much good food is out there that I had no idea I was missing."

JOHN ROBBINS: What rewards can our readers look forward to if they follow your suggestions?

RORY FREEDMAN: Veganism to me is the fountain of youth. When I went vegetarian and then eventually vegan I did it just to be kinder to animals and that was it. But my whole life shifted in so many unexpected ways. I felt physically healthier, I had more energy, and I felt happier. I felt like a more positive human being, and I am sure that can be attributed to the fact that I was no longer eating the fear, grief, sadness, and rage of these poor, tortured animals.

It also gave me the best life because now I have so many amazing, beautiful friends who are also really compassionate and kind. I know a lot of times people will say, "Why don't you care about people too? Why are you only trying to help animals?" I find that funny, because almost all of my friends who are vegan and animal rights activists are also working in other arenas for other causes, whether it's for social justice issues, political issues, or civil rights issues. They care across the board, it's not just, "I only care about animals and I hate people!" They are just compassionate human beings and I get to call these people my friends now because I found them through my participation in the vegan community.

JOHN ROBBINS: I find it a misconception that caring about animals makes you less likely to care about people. In fact I think it starts to orient you towards living and expressing your compassion.

RORY FREEDMAN: I became a much better human being all around when I became vegetarian. That was the beginning of my transformation. Of course I am still transforming and I am still on my path.

JOHN ROBBINS: One critic said that *Skinny Bitch* tells women how to be anorexic. How do you respond to that?

RORY FREEDMAN: I'm not sure what they're using as proof. I mean, in *Skinny Bitch*, we're encouraging people to eat (plant-based versions of) burgers, wings, hot dogs, French fries, pizza, ice cream, and cookies. I have actually gotten an email from somebody who runs an eating disorder clinic. He said, "This book needs to be a part of this community, because I have had so many people who have healed from their eating disorders by reading this book."

It's not my line of expertise by any stretch, but I think that some people who have eating disorders are so afraid of food, they are so afraid of fat, they are so afraid of carbohydrates. They limit themselves to things like five Tic Tacs, four pieces of gum, and coffee all day long. In *Skinny Bitch* over and over again the mantra is, "Eat, eat, eat tons of food. Eat carbs, they are good for you. Eat fruit, it's good for you. Don't be afraid that there is sugar in fruit. It's good sugar. Don't be afraid of nuts and seeds and avocados, those are great fats. Enjoy food, that is what it's there for."

JOHN ROBBINS: Well, we want people to be healthy and we also want them to feel good about themselves. What do you think about the fat acceptance movement? Is it a good thing?

RORY FREEDMAN: Some people bristle about the use of the word skinny, and I get it. For me, that was simply a marketing decision. Women nowadays are obsessed with being thin. That's what a lot of women care about. If you write a book and call it, "*Stop hurting animals. Be vegan!,*" no one will want to buy that book. But if you write a book and call it *Skinny Bitch* then women, as we saw, are going to buy that book by the boatload.

I absolutely think that we need to accept and love ourselves, whatever we look like. That said, when we have a higher Body Mass Index, we are looking at a higher likelihood of cancer, diabetes, and heart disease. These are scientific facts. I don't think we need to be encouraging people who are obese and unhealthy to just say, "This is how I am, and there's nothing I can do about it." I think we all need to love ourselves regardless of how we look, and I think we also need to take better care of ourselves.

JOHN ROBBINS: I am reminded of the Serenity Prayer where we ask for the courage to change the things that we can and the serenity to accept the things that we can't—and the wisdom, most important perhaps, to know the difference.

RORY FREEDMAN: A lot of times people will write and say, "I read your book, it changed my life. I've gone vegetarian. What else can we do?" And I always say, "Whatever it is you were put on this earth to do."

Because we all have a part, whether it's writing books, hosting a conference, writing cookbooks, making documentaries, selling vegan candles or clothes or handbags or purses. Whatever it is that you were put on this earth to do, do it, and do it in alignment with your ethics.

I'm friends with a lot of people in Hollywood who are successful actors and musicians and they were put on this earth to act or to perform, but they also use that platform to promote a more compassionate world. I just think that whatever your walk of life is you can do it with compassion and you can encourage others to join you on that walk.

JOHN ROBBINS: One reviewer called *Skinny Bitch*, "A cynical, foul mouthed read with only good intentions that could get you into your best bikini shape for this pool season." Behind all this attitude, I also find that you are providing solid guidance. What would you say are the central pillars of the lifestyle that you promote?

RORY FREEDMAN: The central pillar is veganism. Then also encouraging people to look at what you are putting into your mouth. Soda for example: What about soda is good for you? Nothing. For some people it could be a huge change if they swap out their sodas and drink water instead. That can actually be a real life changer for a lot of people.

Read the ingredients. Know what it is and ask: "What am I putting in my body?" Also, enjoy food. If you are just starting out on this path, remember I didn't become this healthy whole foods eater overnight. Eighteen years ago, when I first pledged in my apartment in college that I was never going to eat animals again, that was it. I still ate anything I wanted; I just didn't eat cows, chickens, or pigs. I still ate tons of junk food and that's how my journey had to look and that's how I wanted it to look. It took time for it to morph into what it is today. So I always tell people, "Go at the speed that you're comfortable with and transition the way that you're comfortable transitioning."

Some people will say, "Oh, I can't do it. It is too hard. I can't just go from the way I'm eating now to vegan tomorrow." It's okay, you don't have to. Just pick one area to start, and start there.

Steps You Can Take: Action for Humane Food

Visit the Farms

Want to know how animal products are raised? Ask if you can visit farms near you. If you're not allowed to, then you can probably guess that there's something the farm doesn't want you to see.

Visit a Farm Sanctuary

Get to know animals who are being treated kindly at any of the farm sanctuaries located across the United States. For a directory of Farm Sanctuaries, visit *www.sanctuaries.org*.

Want to Include Animal Products in Your Diet?

You can take a step towards a more compassionate world by supporting farms that are giving their animals a decent quality of life. Nothing beats knowing the farm and the farmer yourself, but a third-party verification with a "Certified Humane," "American Humane Certified," or "Animal Welfare Approved" label means the animal probably wasn't raised on a factory farm.

Want Help Going Veg?

For a free vegetarian starter kit, go to *www.goveg.com*, and for a worldwide directory of veg friendly restaurants and grocery stores, visit *www.HappyCow.net*. Also check out *www.greenmenu.org*.

Resources for Ethical Action

Farm Sanctuary

www.farmsanctuary.org

Founded by Gene Baur, Farm Sanctuary combats the abuses of factory farming and encourages a new awareness and understanding about farm animals. Today Farm Sanctuary is the largest and most effective farm animal rescue and protection organization in the United States. Thousands of animals have been rescued and cared for at the three sanctuaries located in Watkins Glen, New York; Orland, California; and Los Angeles, California. The Farm Sanctuary website provides education about the unhealthy and detrimental effects of factory farming on people and the environment, and provides ways to get involved through volunteering and animal welfare legislation. They also offer two guides to vegetarian living.

Righteous Porkchop

www.righteousporkchop.com

This is the website for the groundbreaking book by Nicolette Hahn Niman. She shares her journey to finding a life and good food beyond factory farms.

Humane Farming Association

www.hfa.org

Founded in 1985, and 250,000 members strong, the Humane Farming Association (HFA) is renowned for its highly successful campaigns against animal cruelty and abuse, and for its work to protect the public from the misuse of antibiotics, hormones, and other chemicals used on factory farms. The HFA website offers educational information about the egg industry, the poultry industry, slaughterhouse abuse, the HFA Downed Animal law, and humane education materials.

Humane Society of the United States

www.humanesociety.org

The Humane Society of the United States is the nation's largest animal protection organization, backed by 11 million Americans. It helps animals by advocating for better laws to protect them; conducting campaigns to reform industries; providing animal rescue and emergency response; investigating cases of animal cruelty; and caring for animals through sanctuaries and wildlife rehabilitation centers, emergency shelters, and clinics. The website offers current news and videos on a variety of topics including pet care, animal rescue, wildlife abuse, animal fighting, factory farming, and more. The website provides an adopt-an-animal selection page, and information on animal advocacy, a blog, a state based selection engine, and also supports the *All Animals* magazine.

United Poultry Concerns

www.upc-online.org

United Poultry Concerns is a nonprofit organization located in Machipongo, VA, and dedicated to the compassionate and respectful treatment of chickens, turkeys, ducks, and other domestic fowl in food production, science, education, entertainment, and human companionship situations. The website offers a wealth of information on chickens, turkeys, and other birds raised for food or product consumption.

PART V

The Politics of Dinner—Food Policy for Healthy People

Why are Snickers so much cheaper per calorie than spinach? Why is Coca-Cola often more affordable than clean water? Why are candy bars and cigarettes usually more readily available than fresh fruits and vegetables?

If you want to eat healthfully, why do you have to fight such an uphill battle? Why are government subsidies in so many parts of the world pushing in the wrong direction?

Who would it hurt if we enacted policies that actually encouraged the foods that are healthiest for people and for our world? Who opposes the efforts to make it easier for people to make healthy food choices?

Government policy consistently favors industrial agribusiness, making it harder for small-scale farmers to compete in the marketplace. Agrichemical companies, factory farms, and junk food manufacturers are profiting from a status quo that makes us sick, pollutes the environment, and leaves workers impoverished.

The result: massive illness, trillions of dollars in healthcare expenses, and a whole lot of people going hungry.

But thanks to a passionate resurgence of interest in local foods and food systems, and a shifting economic context, things are beginning to change. Farmers' markets, Community Supported Agriculture, organic farms, community gardens, food-justice organizations, and small-scale family farms are all on the rise. Consumers are taking an interest in knowing where their food comes from, and who grew it. If you want a world where food policy lines up with the health of our bodies, our communities, and our earth, then it's time to get informed, and get activated.

We've got some brilliant experts here to help you do just that.

13

Dr. Raj Patel
Global Hunger, Global Hope

Raj Patel, Ph.D., is a writer, academic, and activist with degrees from Oxford University, the London School of Economics, and Cornell University. He is author of Stuffed and Starved: The Hidden Battle for the World Food System, *and the international bestseller,* The Value of Nothing. *A Visiting Scholar at UC Berkeley's Center for African Studies, and an Honorary Research Fellow at the University of KwaZulu-Natal, in Durban, South Africa, Raj brings an international perspective to our increasingly global food systems. He illumines how your food choices send ripples around the world—and how those ripples come back to impact your life.*

Can we ever end hunger? Yes we can. But it's going to take political will to end poverty, rather than just providing more food in the soup kitchen. What would it be like to participate in a world where everyone can eat healthfully, and live in dignity?

If you care about the social, political, and global impact of your food choices, and you want to know what it will take to create a world where everyone has enough food, then this is a conversation you do not want to miss.

JOHN ROBBINS: One in four children in the United States doesn't know where their next meal is coming from. Many others have health problems made worse by the largely empty calories that are all they feel they can afford. What's it going to take for us to make healthy food available and affordable to everyone?

DR. RAJ PATEL: The reason that half of all American children are expected to at some point in their lives be on food assistance is not because America lacks food. We have more than enough. People who say: "To feed the world we need to produce more food" are barking up the wrong tree. If we can't manage hunger in a country that has more calories available per person than we've ever seen before in human history, then clearly there's something wrong—not with production, but with distribution.

The reason people go hungry in the United States and around the world is because they are too poor to buy food, or to buy healthy food. Tackling this means we're going to have to get a lot more politically active. We're told that when it comes to the food system, we can vote with our forks. That's very well for those of us who can afford to choose what food they put on the end of their forks. But for us to get to a stage where we can consume more wisely, we need to first live in a world where all of us get to consume in the first place.

JOHN ROBBINS: There is an organization that's funded by soft drink companies called Americans Against Food Taxes. They say we can't tax our way to healthier lifestyles. They're obviously a corporate front group, but I agree with them on one point: That if we only discourage unhealthy foods without making it easier and less expensive for people to eat healthier ones, we're not going to succeed. Do you think there are ways that we could tilt the playing field to help people of low income have an easier time making healthier choices?

DR. RAJ PATEL: Merely sticking higher prices on food is going to hurt the poor. But it's absurd for soft drink companies to claim that taxation isn't going to solve anything. According to data from the University of California at San Francisco, a penny-per-ounce tax on sweetened beverages could prevent nearly 100,000 cases of heart disease, 8,000 strokes, and 26,000 deaths over the next decade.

When you look at the actual data, instead of listening to the whining of the soft drink companies, the need to shift our consumption away from foods that are high in sugar becomes clear.

For an example of how this can be done, let's think about tobacco. When it became increasingly clear that tobacco needed to be regulated, the approach wasn't simply to tax it. We also restricted tobacco advertising, provided education on the effects of tobacco and how to quit, funded more public health research and, most important, took on the tobacco companies themselves.

We can apply the lessons we've learned from tobacco policy to junk food. We should also be asking some fairly revolutionary questions, like:

1. Why is it that we allow corporations whose products are responsible for a great deal of ill health to carry on in the way that they do?

2. People should be free to consume what they like. But why is the playing field of consumption so heavily tilted towards the most powerful corporations and away from the kind of sustainable diet that's good for the environment, jobs, individual bodies, and a healthy culture?

3. Why have fresh fruits and vegetables become more expensive in the United States in recent years, while processed food has become cheaper?

Certain corporations bemoan the idea of taxes on junk food, when, in fact, a great deal of our taxes currently go to supporting the production system from which these same corporations profit. Having cheap corn, for example, is a great benefit to the manufacturers of high-fructose corn syrup, and also to the producers of ethanol and biofuels. Having these kinds of subsidies in place at the expense of a sustainable food system is absurd.

JOHN ROBBINS: Our subsidy of corn is also of great value to the livestock industry, and in particular to reducing the price of factory-farmed meat, because a great deal of the corn grown in the United States is fed to livestock.

DR. RAJ PATEL: And another reason that factory farming is profitable is because the industry doesn't have to pay the long-term costs of the pollution they produce. The runoff from concentrated animal feeding operations flows into the Mississippi River, for example, and it's created a huge dead zone in the Gulf of Mexico. The costs of that are immense, but they are not borne by the operations that cause it. There are substantial environmental and social costs associated with the production of meat today. The industry as it is currently established is able to be profitable precisely because it escapes the consequences of its actions. They argue that what we need in order to feed people in the world is cheap food. But I say that what we really need isn't more cheap food—we need higher wages. We need people to be able to afford to buy food that pays fully for its environmental impact, and pays the workers who produce it.

JOHN ROBBINS: In *Tomatoland*, Barry Estabrook describes how he sought to discover why the fresh tomatoes that are available at most stores today are so tasteless, and he wound up uncovering deplorable working conditions for most of the laborers. He even found people chained inside their living quarters to prevent them from running away, which is a form of slavery. He found workers being forced to buy their food and supplies from company stores at grossly inflated prices, and that they were not making anything close to a livable wage.

In 2012 Trader Joe's finally relented after years of protest and signed an agreement with the Coalition of the Immokalee Workers, agreeing to pay a penny more per pound of tomatoes. Whole Foods, Taco Bell, McDonald's, and Burger King had all signed the agreement years ago. Finally Trader Joe's has joined them. How did you feel when Trader Joe's signed the agreement?

DR. RAJ PATEL: I tweeted the word "Victory!" because I was very pleased that a grassroots campaign, led by some of the poorest people in the United States, the Coalition of Immokalee Workers, had brought this industrial food giant to the bargaining table. I think the Coalition was right to celebrate that night, and then right to get up in the morning and carry on working. Before the agreement, the average wage of tomato pickers, who lifted 2.4 tons of tomatoes every day, was about $11,000 a year. After the wage increase, the average is now closer to a whopping $20,000 a year. Obviously, no one is going to be living in the lap of luxury on $20,000 a year in the United States.

The retailers that command our food systems have moved the needle from intolerable working conditions to barely tolerable, but the situation is still bad. This doesn't represent a revolution in food, but it is a positive step.

The good news is that more and more people are listening and trying to amplify the voices of the workers themselves. We're seeing increasing numbers of workers' groups and communities of people who are directly affected by the food system getting together and organizing.

If we are concerned about making sure that everyone gets to eat well and have jobs that are dignified and well paid, that's going to take more than the gift of a retailer or two deciding to make things a little better for the worst paid people in the country. It's going to require sustained local organizing from broad coalitions. The efforts of the Coalition of Immokalee Workers were supported by faith-based groups, student groups, journalists, and unions. All these groups got together and went to Congress, and to these unaccountable growers, demanding changes and ultimately winning a victory.

JOHN ROBBINS: Why do you think it was that Taco Bell, McDonald's, and Burger King signed the agreement, and yet Trader Joe's, which presents itself as a more progressive, health-oriented, and sustainable seller of foods, was so slow to do so?

DR. RAJ PATEL: I think it's a reminder that if we see green marketing materials in some of the groovy and granola-rich places we get our food, it's often just window dressing. Those of us who like to eat sustainable, local, and organic food need to listen also for the voices of the workers. In California a couple of years back we had a woman die in the fields of an organic food producer because they weren't allowing her to have access to water in hundred-degree-plus weather. So although the label might have had a happy smiling farmer and the USDA organic certified logo, and although it was— for Californians—locally produced, the fact is that workers died in the making of that food, right in the United States.

If you're serious about a revolutionary food system, you have to go beyond the individual model of consumption where you read the label and feel good about yourself. You have to work with the entirety of the food system. Invariably that means labor, and if your food is grown in the United States, that means people of color. When you see the marketing of green and happy but you don't see people in the fields, then chances are, the people you don't see are going to be people of color who are being treated badly.

JOHN ROBBINS: The entirety of the food system also includes companies like Goldman Sachs and other big players in food commodity trading. They are today making billions of dollars at the expense of millions of people in poor countries. A few years ago they were profiting off of subprime mortgages, and now they're profiting off the price of food commodities. In each case, investors leveraged hundreds of billions of dollars for short-term profits, in spite of the consequences. What do you see as the role of hedge funds and investment banks in the rise of global food prices?

DR. RAJ PATEL: In 2008, the price of gas shot up suddenly. This was driven in no small part by traders on Wall Street betting that the price of oil was going to keep going up until it reached $140 a barrel. Then the bubble burst and the traders who were holding contracts for oil at ridiculously high prices realized that they needed to cash out, and wanted somewhere else to put their money. They made the cold and hard calculation that even if the price of gas was on its way down, the one thing that folks always need is food. To some extent the cash that was left over after the housing and oil bubbles burst, went into food. Increased speculation drove up the traded price of food. On top of speculators on Wall Street betting on the price of food and driving up prices, many retailers took advantage of the fact that prices were going up to add a little bit more onto their markup.

The price of food has some very important consequences. January of 2011 was when what became known as the Arab Spring started. It actually started in Algeria, with a rise in the price of wheat. Because people couldn't afford wheat, and because the government was unwilling to do anything about it, people took to the street. Now in Algeria the government successfully beat down the uprisings. But those uprisings ended up spreading throughout the Middle East and across the Mediterranean. Although food prices were not the whole issue, they have had an important impact in terms of political stability around the world. Speculators were fueled by greed to bet on rising prices, governments did nothing about it, and the people took to the streets. In a sense, the gougers and the profiteers on Wall Street were hastening a rise in political engagement.

JOHN ROBBINS: About a billion people globally go hungry every day. In a bizarre mirror image, there are about one billion people on earth today suffering from diseases caused by eating too much fat, too many calories, too much sugar, and too much junk. What do you see as the connection between the billion-plus of us who are stuffed, and the billion who are starving?

DR. RAJ PATEL: The connection is poverty. If you're living in the United States and you're getting supplemental nutritional assistance, which used to be called food stamps, then you'll get about $1.50 to buy a meal. Now $1 will buy you 440 calories of cola, but it will only buy you 17 calories of lettuce. The rules of the game suggest that if you're just trying to make it through the day with enough to eat, you're going to go for the calories that you can afford. Again, this isn't to say that you necessarily eat badly. There is plenty of evidence that shows how people on reduced incomes are surviving in incredibly creative ways. But the rules of the game are tilted against people on restricted incomes.

JOHN ROBBINS: Do you think it's possible we will ever be able to eradicate hunger from our world?

DR. RAJ PATEL: Yes, if we can end poverty. And we've already seen some tremendous advances. In the United States, the war on poverty did a great deal to reduce the levels of hunger. It's only been in the era of freewheeling, free markets that we've seen the number of hungry people skyrocket from 20 million to now 50 million.

We can end hunger, but it's going to take the political will to end poverty, rather than just increasing assistance to people on the breadline. We need to work for a world in which everyone gets to eat and everyone can live in dignity.

The only thing standing between us and a world free from starvation is political will and our own imaginations.

14

Michele Simon

How the Food Industry Undermines Your Health and How to Fight Back

Michele Simon is a leading expert on the politics and practices of the food industry. She is author of Appetite for Profit, *and president of Eat Drink Politics, an industry watchdog consulting business. As a public health lawyer, Michele has done groundbreaking work on issues like food safety, agricultural policy, health claims, nutritional labeling, school food policy, and consumer empowerment. In an industry often dominated by big money, Michele is a champion of the public welfare.*

The companies behind foods that make people fat and sick claim to care about kids. But they are actually lobbying to make sure they keep their highly processed junk food products in front of kids at all costs. How does the food industry dominate public policy, and what can you do to protect your family and to stand up for the common good?

JOHN ROBBINS: In 2012 Congress decided that pizza sauce could be allowed to meet the "vegetable" requirement that had been established for the national school lunch program. How does the processed food industry get away with stuff like this?

MICHELE SIMON: We have a long history of food industry influence and control over the school meal program. It's no secret that the meals served in most schools in the United States are a nutritional disaster. There are some good projects going on, like salad bars and school gardens. But for the most part, school meals remain extremely unhealthy. They are packed with additives, processed food, dyes, animal products, and sugar.

The U.S. Department of Agriculture (USDA) has authority over school meals, and they attempted to put together some reasonable and improved nutritional guidelines. This wasn't something that just came out of the Obama administration or the First Lady's Let's Move! program. It had been under development for years.

Science points in the direction of more plant-based foods and less processed foods, so that is what the USDA was trying to accomplish. The rule is that two tablespoons counts as a vegetable serving, which is obviously something that serves the frozen pizza industry very well, but it doesn't really help kids be healthier.

The USDA took a look at the regulation that considered pizza sauce to be a vegetable, and was trying to change it. They were trying to make other improvements as well, like taking Tater Tots and French fries off the list of everyday menu items.

These attempts to improve school meals didn't go over well with either the frozen pizza industry or the potato industry. So they sent their lobbyists to Congress to undermine the USDA's best effort at improving school meals. Members of Congress, unfortunately, do the bidding of those who pay for their reelection campaigns. So the result of business as usual in Washington, D.C., is that pizza sauce continues to be considered a vegetable, and Tater Tots and French fries will remain on school menus far too often. Food industry lobbyists won, and kids lost.

Despite the setback, the USDA did manage to instigate some improvements to the school meal guidelines, including additional requirements for fruits and vegetables to be served.

JOHN ROBBINS: More and more people are aware of how they eat and want to make better choices for themselves and their families. People are cutting down on animal products, eating more whole grains, eating more real food and less processed food, buying more locally, and making thoughtful choices about whether or not to consume GMOs. Those are all things you support, and yet you also remind us that personal action alone is not enough.

MICHELE SIMON: Too many of us still do not have access to healthy, affordable, and fresh food. That's where the politics come into play. Once you've looked around your own kitchen cupboards, then it's time to look beyond your home and neighborhood. Chances are, you don't live far from people who don't have access to healthy food that they can afford. I feel like people who are fortunate and privileged to be able to make healthy food choices have a responsibility to look out for the well-being of those who do not have that opportunity.

There are a number of ways to get involved. You can work to improve school meals in your local school district, or help local schools to start school gardens. You can help start new farmers' markets. You can help regional food programs or food banks improve the healthfulness of the foods they offer to poor neighborhoods. You can lobby at the mayoral, state, or federal level. There is no shortage of opportunities to get involved. The key thing is to find some way of getting engaged that speaks to you and sparks your interest.

JOHN ROBBINS: It's commonly believed that we are each responsible for what we put in our mouths, and therefore if people make poor food choices, it's their own fault. But you point out that this is only a half-truth, because food industry marketing bears much of the blame for the health crisis. Do you think it would be accurate to say that the food industry today is actually undermining our health?

MICHELE SIMON: Yes. It's not that the CEOs of Kraft and McDonald's and Coca-Cola are sitting around trying to figure out how to make people unhealthy. What they are doing is sitting around trying to figure out how to sell more products. It just so happens that processed foods really are the least healthy products for us. The name of the game in capitalism is growth, and food companies have to grow or they die. The only way to grow is to keep selling more products. Once they saturate one geographic market, they have to go to the next geographic market, which is why all of the major food companies are multinational in scope. They're not just ruining America's health now. They're going overseas to ruin the health of Europeans and, increasingly, the citizens of developing nations who want to be like Americans. So while they're trying to shop like Americans and eat processed foods and animal products like Americans, they're now going to get sick like Americans.

JOHN ROBBINS: There are actually more Baskin-Robbins stores now in Tokyo than there are in Los Angeles.

MICHELE SIMON: The food industry's job is to sell more products. It's not their fault; it's just what they're in business to do. The problem I have is when they are deceptive about it, or when a company like McDonald's claims to care about kids and to be making changes to their Happy Meals, when it's really just a lot of deceptive window dressing.

We have a number of products on the market that are trying to claim to be healthy because the food industry hears the human cry. They realize that people are suffering from a number of diet-related diseases in which they have a role, and have decided that they have a public relations problem on their hands. They often try to spin their products as good for you. Companies like Kraft and PepsiCo are masters at this. They come up with healthier Cheetos and diet sodas that are supposedly better for you. They look for any way they can find to keep America eating their highly processed, unhealthy junk food.

At the same time that these companies are claiming to be a part of the solution and improving their products, behind the scenes it's business as usual. The same companies that claim to care about kids are actually lobbying to keep the status quo and to make sure their highly processed, unhealthy foods are in front of kids as much as possible.

JOHN ROBBINS: Wal-Mart now sells 25 percent of the groceries purchased in the United States. Most of those sales are in rural areas, but the company has said that it intends to expand into urban markets like New York. What do you think will happen if the company succeeds in penetrating urban markets?

MICHELE SIMON: We know from Wal-Mart's march across rural America that the economic impacts of this monster retailer have been devastating. It's led to the closing of mom-and-pop stores all over the country that couldn't compete with Wal-Mart.

We already have some evidence of what would happen in urban areas, because Wal-Mart opened several stores in Chicago. There was a study that looked at the impact on surrounding food retailers. Sure enough, just like they put five-and-dime stores out of business across rural America, they're putting food retailers out of business in Chicago, too.

Wal-Mart always comes in claiming to be a savior by offering lots of new jobs. And they do create some jobs, but the evidence shows that they destroy more jobs than they create. And the jobs that they do create offer extremely low pay. Wal-Mart is notorious for their labor violations. For example, they are extremely hostile to unions. Any place where a union has attempted to be formed in a Wal-Mart, the company responds by shutting down the entire store. Many Wal-Mart workers depend on food stamps to survive. Federal, state, and local governments have to subsidize Wal-Mart's workers because they're being paid so little. That has an impact on every taxpayer.

Then there's Wal-Mart's impact on the food supply. They claim that they'll try to source local food, but we know that Wal-Mart's business model is to squeeze the supplier. Farmers are already extremely underpaid. Small farmers are becoming a dying breed in America. To have Wal-Mart's business model applied to local produce will mean small farmers not getting paid enough to survive.

The good news is people in New York are wise to this and many of them don't want Wal-Mart in their communities. There's a huge debate going on, and Wal-Mart's track record is increasingly well-known.

JOHN ROBBINS: When corporations dominate the conversation, the truth is often treated like a necessary casualty in pursuit of the bottom line. One example is the meat industry, which has been working overtime for decades, trying to convince people that we need meat to get enough protein.

MICHELE SIMON: The meat industry recognized decades ago that they could strengthen their market share in the American diet by promoting the notion that protein equals meat, and they have been very successful. Meat industry trade publications use the words protein and meat interchangeably, which should give us an idea as to where this brainwashing really comes from.

We've known since the 1950s that too much meat in the diet can cause heart disease, colon cancer, and a number of chronic illnesses. The meat industry realized it had a problem on its hands, so it's been working hard to keep the emphasis on protein. While protein is certainly a nutrient we need, the industry has been using it to distract attention from the less desirable things in meat—like fat, cholesterol, hormones, and all the chemicals that go into processed meats.

JOHN ROBBINS: In 2011, the Environmental Working Group released a report on sugar in children's cereals. They found that, not too surprisingly, many popular brands package a tremendous amount of sugar. Some very popular cereals are actually higher in sugar than Chips Ahoy! cookies. They also found that the food industry continues to market these products aggressively to children.

MICHELE SIMON: I believe that marketing to children is immoral as well as illegal. Because the issue of marketing to children has been coming under scrutiny, the food industry has claimed to have changed its ways and become self-regulating. The Environmental Working Group's report showed that the industry wasn't even following its own very lame guidelines.

JOHN ROBBINS: I remember seeing a child wearing a T-shirt that said, "If you love me, don't feed me junk food." And here we have corporations with very sophisticated marketing campaigns to which they devote tremendous financial resources, preying on children's formative psyches. These kids are developing patterns and habits of food consumption that will presumably be lifelong. It seems like food corporations have become masters at invading our consciousness and getting us to eat what we shouldn't eat, to eat more than we should, and to vote for people who serve their financial interests without much regard for our health. Yet most Americans are too busy making a living and dealing with the pressures in their lives to muster the energy that would be needed to rebel effectively against this indoctrination. What are some of the most effective things committed and informed people can do?

MICHELE SIMON: The food industry has way more power than it ought to have in Washington and in statehouses, but we have people power, the truth, and morality on our side. A movement is growing. You can see it not just in the growth of farmers' markets and natural foods, but also in efforts from the USDA and government agencies to make positive changes, and in all the local programs, nonprofits, and for-profit companies that are picking up on these messages and trying to make a difference. There is an explosion of energy to try to fix what more and more people realize is a major problem.

If you care about food, it's important to connect on the political and systemic level. Look for ways to support groups that are engaged with the Farm Bill, school lunches, GMO labeling, or so many other areas of importance. Don't let your country's food policy be a spectator sport.

15

Dennis & Elizabeth Kucinich
Bringing Sanity to Public Food Policy

Elizabeth and former U.S. Congressman Dennis Kucinich are leading champions of efforts to bring sanity and health to public food policy. Elizabeth Kucinich is the director of government affairs for the Physicians Committee for Responsible Medicine. Her husband, eight-term U.S. Congressman and former Cleveland Mayor Dennis Kucinich, has run for president twice and been a passionate spokesperson for the regulation of GMOs and for healthier school lunches, along with many other causes. Dennis and Elizabeth Kucinich work in the heart of politics, and they bring compelling insights.

How have monopolies in agriculture and banking destroyed family farmers and made America a less democratic nation? What's really fueling the crazy food policies coming out of Washington, D.C., and many other nations—and what can you do about it?

JOHN ROBBINS: Polls show that about 90 percent of the American public supports the labeling of genetically engineered foods. Dennis, you have many times introduced legislation in the House that would require genetically engineered foods to be labeled, but none of these bills have ever passed. In 2012 Monsanto actually threatened to sue the State of Vermont if legislators there passed a GMO labeling bill. What is at stake in these battles and why have you been so persistent in seeking labeling?

DENNIS KUCINICH: What is at stake is our liberty. When you have Monsanto threatening the government of the state of Vermont, then you know that Monsanto has become too big. Then you know that they are subverting the political process to deny the American people the right to know.

It's a kind of corruption that exists at every level of government when corporations have dominant influence over public policy. They are going to ruin public health unless they are held to a higher standard, unless there is labeling, unless we decide that people do have a right to know.

There haven't been any longitudinal studies on human health done on the consumption of genetically engineered products, and yet we know that these products do have affects. People should have a right to understand if there are implications with respect to allergenicity, toxicity, and antibiotic resistance. These are all things that a Genetically Engineered Food Right to Know Act would take into account.

JOHN ROBBINS: Elizabeth, you are a co-producer of *The GMO Film Project*, which tells a remarkable story of poor Haitian farmers rejecting Monsanto's help after the devastating Haitian earthquake. What happened?

ELIZABETH KUCINICH: Immediately after the 2010 earthquake destroyed so much of Haiti, Monsanto offered to donate 475 tons of genetically engineered seeds. The peasant farmers' movement got together and 10,000 people demonstrated in the streets, where they symbolically burned the seeds in protest. They understood that planting those seeds would have had profound impact on their food sovereignty, and they were also aware that these crops would require them to use pesticides and artificial fertilizers. We can learn from these poor Haitian farmers, and be inspired by the stand they took.

JOHN ROBBINS: In the United States, the federal government subsidizes genetically engineered feed grains—mainly corn and soy, which are fed to livestock in factory farms. In fact, almost two-thirds of federal subsidy funding currently ends up supporting meat and dairy products, and a great deal also ends up lowering the price of things like high-fructose corn syrup. Less than 1 percent goes to fruits and vegetables. At a time when we have the highest rates of obesity of any country in the world, and when one in three U.S. children is expected to get diabetes, we subsidize the foods that make us sick.

ELIZABETH KUCINICH: Physicians Committee for Responsible Medicine (PCRM) did a report on the Farm Bill and where the billions of dollars in subsidies actually ended up. The report sent a shock wave through Congress. People started asking: Where are we putting our money? Shouldn't we be supporting outcomes that we actually want to achieve, as opposed to making our nation more unhealthy?

A lot of members of Congress have not been on a farm and they haven't ever studied nutrition. It is up to us to engage and inform them. When your representatives hear from you, you get an opportunity to influence their education and their thinking.

JOHN ROBBINS: Four firms control more than 80 percent of the beef slaughtered in the United States. Is there any possibility of using antitrust legislation to break up monopolies that companies like Tyson, ConAgra, and Cargill have developed?

DENNIS KUCINICH: We need a serious discussion in America about the impact of monopolies in agriculture. Horizontal and vertical monopolies have created an agribusiness monolith that includes pharmaceuticals, the chemical industry, the biotech industry, and the banks that back them up.

As we have learned, when banks are too big to fail, they can bring down our economy and represent a threat to our freedom. Monopolies in agriculture and banking have destroyed family farmers, and have actually made America a less democratic nation. We need to stop the vertical and horizontal integration that has gone on financially, to the detriment of a lot of small producers. We need to help people go back to the farms, and we need to put limits on the amount of farmland that any one company can own.

JOHN ROBBINS: It seems as though the agrichemical companies and the factory farms have created an artificial distortion in the marketplace. They have been able to externalize the true costs of their products onto the rest of us and onto future generations. They have been able to keep authorities from enforcing antitrust legislation and from enforcing pollution regulations. They have been able to profit at the expense of the health of consumers, our communities, and our environment. What is it going to take to turn this around so that healthy, earth-friendly, and sustainable agriculture can prevail?

DENNIS KUCINICH: The policy change that is needed goes to the core of whether we even have a republic or a democracy. The Supreme Court ruled in the Citizens United case to let corporations directly influence federal elections. They allowed government to become an auction, and policy goes to the highest bidder. Monopolies, by virtue of their economic concentration, have the ability to bid higher than anyone. For example, 90 percent of the American people want to label genetically engineered food products, and yet we don't have labeling. Monopolies have suffocated our ability to be a government of the people, by the people, and for the people.

So what is it going to take? It is going to take raising consciousness about the condition, it's going to take people becoming visibly engaged in the political process, and it's going to take a mass movement. But it really begins with education, which has the ability to be the spark that ignites the movement.

JOHN ROBBINS: Dennis, you've been working valiantly for many years to build this movement. And you seem to have boundless energy. You are, as I am, 65 years of age. Elizabeth is 34. She is 31 years younger than you and has tremendous energy. Do you feel that your healthy vegan diet is helping you to keep up with her?

ELIZABETH KUCINICH: It's the other way around. You try chasing after Dennis in Congress!

DENNIS KUCINICH: In 1995, I switched to a vegan diet after having spent most of my life dealing with a very crippling case of Crohn's disease. I had been hospitalized repeatedly, and been brought to death's door when I was 21. I had not connected the food that I was eating to the medical condition that I had. But when I changed my diet and stopped eating processed foods, animal products, and sugar, the Crohn's symptoms disappeared. I had a new level of vitality and energy. Becoming a vegan not only saved my life, but it probably added another sixty years to it.

JOHN ROBBINS: And Elizabeth, how did you become a food activist?

ELIZABETH KUCINICH: My mother also had Crohn's disease, and she almost died when I was five. The doctors didn't know what they could do to help her, and she just said: "That's it. Forget hospitals. I am going to study nutrition." She studied whole foods living, and wound up eliminating meat and dairy from her diet. She transformed herself, and today she is a wonderful, healthy lady.

When I saw my mother undergo this amazing transformation just through changing the way that she ate, I started looking into how food was produced. What I learned led me to become an activist.

Everybody has their own journey, but very often health is a crucial doorway. You can take drugs and you can have surgery, but what if you look at the really simple building block of life, which is what you are using to build and fuel your body?

I am thrilled now to work with the Physicians Committee for Responsible Medicine in Washington, D.C. It is, to me, a sanctuary in the capital. We've got eighty people in a building who are all working towards a world of greater health, sustainability, and compassion.

JOHN ROBBINS: Have you or PCRM looked at how much money we would save treating disease in a country if we were to prevent a great deal of it in the first place by supporting healthy nutrition policies?

ELIZABETH KUCINICH: We could save many hundreds of billions of dollars a year. Research indicates that diabetes treatment alone will cost more than $3 trillion in the

United States over the next decade. This is largely preventable. There is so much we can do.

JOHN ROBBINS: I want to thank you both. As a partnership you embody the male and the female working together in purpose and harmony for a higher vision. Dennis, over your eight terms of congressional service, and through two Presidential campaigns, I believe you have been one of the great leaders in American public life. I believe every citizen, and the very cause of democracy itself, owes you a debt of gratitude for your service. Now it's up to all of us to carry the work forward.

16

Morgan Spurlock
Super Size Me

Morgan Spurlock knew that fast food was cheap, convenient, and tasty, but that eating it could lead to obesity and higher rates of heart disease. He wanted to find out: is it really all that bad? Or is it worse than you ever imagined? He came up with a bold experiment, and decided to share it with the world.

In Morgan's film, Super Size Me, *he ate nothing but McDonald's meals for thirty days, and tracked the results with a team of doctors.* Super Size Me *premiered at the Sundance Film Festival in 2004 and won Morgan Best Directing honors. The film went on to win the inaugural Writers Guild of America Best Documentary Screenplay Award as well as garner an Academy Award nomination for Best Feature Documentary. Morgan then became the executive producer and star of the reality TV series,* 30 Days, *and many other programs.*

What did Morgan find out in his bold McDonald's experiment? It just might change your relationship to fast food . . . forever.

JOHN ROBBINS: I remember talking with you about the idea of *Super Size Me* before you made it. I felt you were going to create something extraordinarily powerful.

MORGAN SPURLOCK: Or die in the process.

JOHN ROBBINS: What inspired you to do it?

MORGAN SPURLOCK: When I got the idea for *Super Size Me*, I was at my mom's house in West Virginia, where I grew up. I was on her couch in a tryptophan haze on Thanksgiving Day. There was a news story on about two girls who were suing McDonald's. They were basically saying, "We are sick. We're obese. And it's your fault." That was the argument their lawyer was making. I thought, "Well that seems a little crazy. They are just going to blame one place for all of their problems? What else went into this? Could it just be McDonald's fault?" Then McDonald's came on television and said, "Listen, you can't link our food to these girls being sick. You can't link our food to these girls being obese. Our food is healthy. It's nutritious. It's good for you." I was like, "Well, I think you're getting a little carried away too, because if that's good for me, shouldn't I be able to eat it for thirty days straight with no side effects?" The light went on and I was like, "I just got an idea for a movie." That's literally where it came from.

JOHN ROBBINS: I view what you did in eating that food exclusively for a month in a context similar to Dr. Martin Luther King marching and being arrested and undergoing suffering on behalf of a greater cause.

MORGAN SPURLOCK: He was arrested. I was almost cardiac arrested.

JOHN ROBBINS: You put yourself biologically and cellularly through some serious consequences. What surprised you most about what took place?

MORGAN SPURLOCK: I was surprised by the speed at which my body started to collapse, and the things that were happening within me. I was not a vegan when I started this, nor even a vegetarian. I was eating just kind of the traditional American diet. During the thirty days, my liver filled up so quickly and was just filtering through so much fat, that Dr. Isaacs famously said: "Your liver is turning into pâté." Both he and Dr. Lisa Ganjhu, who is the gastroenterologist I was seeing, were really shocked by the damage that this experiment was doing to my liver.

JOHN ROBBINS: Your liver got so fatty that the enzymes were off the charts.

MORGAN SPURLOCK: Yes, and I was really lethargic. I felt terrible. I was so run down. I wish I had monitored the amount of sugar that was in my body over the course of that diet, because I was having these incredible highs and lows and I'm sure it would have been incredible to see.

JOHN ROBBINS: You know there are people in our society whose energy and mood swings are so severe that it's almost as if they are bipolar. It's not that they actually have a biochemical mental illness. It's that their blood sugar levels are gyrating and don't have a sense of balance.

MORGAN SPURLOCK: I can even see it in my five-year-old boy. If he just has a little bit of sugar, what happens to that kid is amazing. In less than a minute after it gets into his bloodstream, he's bouncing off the walls. He is running around, and then just as quickly his energy crashes. It's amazing to see.

JOHN ROBBINS: Was there a backlash from McDonald's or from other fast-food companies?

MORGAN SPURLOCK: McDonald's was really smart about how they handled the film in the United States. When Eric Schlosser wrote *Fast Food Nation*, McDonald's aggressively attacked him and his book. I think that they saw how that tactic blew up in their face, and how it really didn't gain them any new fans. So when my film came out in the United States they said, "You know, we're just going to ignore this. He's crazy. He was overeating and under exercising. So what does he expect?" Forty-five million people go to McDonald's every day around the world. Half of that number, more than 22 million, are in the United States. Rather than jeopardize half of their business, they said, "We'll just ignore it in the United States." I think it was a wise move on their part.

What happened in other countries, though, was incredible, because they attacked the movie viciously internationally. They took out full-page ads in newspapers. I would go on the news and the CEO of McDonald's would come on right after me to talk about how this is untrue, or it didn't happen. In Australia, right before the film opened, the last movie trailer you would see was Guy Russo, the CEO of McDonald's, sitting on a fence in blue jeans, a blue shirt, and a cowboy hat going, "G'day mate. The film you're about to see is a lie." It was incredible. They really went after the movie. The greatest advertising we had was their attack campaign.

Right when the film came out, they started giving out apples at McDonald's. That was the big revelation that they had. And so McDonald's called the owner of one theater chain in Australia and said, "We want come down to the theater the day that *Super Size Me* is playing, and when people are coming out of the theater we want to give everybody who is coming out of the movie theater apples. What do you think about that?" The owner of the theater then proceeded to tell McDonald's what they could do with their apples.

JOHN ROBBINS: You know when they post on their signs how many billions they've sold, they're not talking about apples. They're talking about burgers.

MORGAN SPURLOCK: That is exactly right.

JOHN ROBBINS: And a high percentage of their 22 million daily U.S. customers are overweight or even obese. We in the United States have the highest rates of obesity and morbid obesity in the history of the industrialized world. And of course this is not just an issue of aesthetics and vanity. There are very real and serious health implications. One of the themes that I picked up from *Super Size Me* was that fast-food companies—McDonald's in particular—actually have ways of encouraging people to overeat or to eat badly because that's how their profit increases.

MORGAN SPURLOCK: The whole idea of supersizing and biggie sizing and king sizing, as it was used by McDonald's, Wendy's, and Burger King, originated with movie theaters. They saw the margins and that they could double or triple the price of a popcorn as it went from one size to another. The added popcorn cost them like a nickel more to make, and they were able to make a tremendous amount more money. They doubled the profit margin on things like this. So the fast-food restaurants said, "There's a real business model that we should tap into." By giving more fries, by giving more soda, they could add fifty cents to the price and pay a nickel for the added food—and it was good for their profit margins.

JOHN ROBBINS: We have a very skewed system where the calories that are least expensive for people to buy are often the least healthy. What you have to pay to get 500 calories worth of broccoli is a lot more than what you pay to get 500 calories worth of soda pop or cheeseburger.

MORGAN SPURLOCK: When you go to a grocery story, it's often the things that are the best for you that are the most expensive. When you're looking at junk food, you can go over to the aisle where everything is subsidized through government funding, and you have these things that are just loaded with corn syrup, soy lecithin, and all sorts of additives. You have cookies that can basically live on the shelf for years and look the same. It's like the only thing that will be left after a nuclear war would be these cookies and a cockroach.

But you go over to the other aisle and here's all of this incredibly healthy food that certain parts of our society couldn't even afford because of how pricey it is. I feel that we have a really skewed vision of our health in the United States. We just don't put a value on it, and the value we put on food is a profit margin rather than what's good for us.

JOHN ROBBINS: Corn and soy are heavily subsidized products. They are also the two products that by far are the most heavily genetically engineered, and they provide the basis of most livestock feed in the United States. In fact, most of our corn that isn't going to biofuels is going to feed animals—particularly hogs, beef cattle, and dairy cows, as well as poultry. The same is true for our soy crop. Very little of it is going to tofu. It's going to livestock feed. By subsidizing soy and corn like we do, what we're really subsidizing is factory-farmed meat and high-fructose corn syrup.

MORGAN SPURLOCK: What do you think is the biggest thing that's keeping these misguided subsidies in place? Is it just the big-business lobby? Is it all the money that's basically pumping into our system in D.C., making sure that no one is ever voting for what's best for a carrot?

JOHN ROBBINS: It's big agribusiness and big lobbying dollars from institutions like the Sugar Association, The National Livestock and Meat Board, National Beef Council, National Dairy Council, The Milk Board, and the Grocery Manufacturers Association. All of these groups have so much power and influence over the Farm Bill, and over any kind of legislation that takes place.

MORGAN SPURLOCK: There's no one really being a big voice for what is good for you.

JOHN ROBBINS: Some people do take a stand, even if they don't have a lot of resources to lobby politicians. Corporate Accountability International has been running full-page ads in major newspapers calling on McDonald's to retire Ronald McDonald, the character who's been the face of the company. In view of the escalating rates of childhood obesity and diabetes, they are basically saying, "Stop marketing to kids," "Stop putting a clown there as if what you're selling is fun."

MORGAN SPURLOCK: I remember when *Super Size Me* came out. Childhood obesity was very much on the rise, as it still is today. McDonald's had a press conference and they said, "We understand. We really understand what a problem this is and we want to play our part to help combat childhood obesity. That's why McDonald's, with the help of Ronald McDonald, is going to make an exercise video for kids." And so Ronald McDonald basically made an exercise video with all of his friends. They said they wanted to combat obesity, but they didn't change their menu.

JOHN ROBBINS: Some people try to raise their kids to eat healthfully, but struggle with the three-headed monster of peer pressure, advertisements, and the fact that a lot of junk food is designed to be addictive. The more you have, the more you want.

MORGAN SPURLOCK: My little boy has never had any of that fast-food type of junk food so he doesn't really have a taste for it. He doesn't even want it. He didn't eat a lot sweets growing up, so he doesn't really have a taste for them, either. When he gets them he's happy to have one, but he doesn't even eat the whole thing. For me it's interesting to see how that has shaped what he likes to eat. You give that kid cucumbers and he'll chow them down. I think you start to model after what you see your parents eat, and what you're used to. Now that he's in kindergarten and peer pressure is starting to kick in, he sometimes talks about what the other kids will have for lunch. But he never really asks for it. He knows that these are things that he just doesn't eat.

JOHN ROBBINS: You have been able to start him out from the beginning with a really healthy foundation. It will pay dividends all the way through.

MORGAN SPURLOCK: And then he'll hate me later.

JOHN ROBBINS: I think he'll love you later because most people do. And one of the things people love you for, in addition to *Super Size Me*, is your show *30 Days*. If our job on this earth is to overcome intolerance and develop empathy and compassion, I think *30 Days* might be the best television series ever made. Each episode follows someone through 30 days of doing something way outside their comfort zone, which gives them empathy for a totally different kind of life experience than what they've ever known.

I remember one of the episodes in which there was an avid hunter named George from North Carolina. He lived for 30 days with a People for the Ethical Treatment of Animals (PETA) campaign coordinator and her vegan family in Los Angeles. George participated with his host in PETA projects. He worked at an animal rescue center. He ate no meat for the 30 days. He took part in animal rights demonstrations—he even wore a big chicken costume in one of them. I met George at the end of his *30 Days* and he seemed to be deeply changed in a very positive way by the experience.

MORGAN SPURLOCK: What happened with George over the course of that show was really amazing. He came in as a lot of people do, having a specific idea of where his food came from and what it meant. You could actually see him having an awakening over the course of this show, as he started to understand the impact of factory farming and what it means to us and to our society. They went and rescued a downed cow that was basically just tossed away, and you could see him nurture this cow back to health. He was so impacted by this. I love to see this type of transformation. What I loved about the show was that there were these real awakenings over the course of that series where people would suddenly be like, "Oh my god. I get it now."

JOHN ROBBINS: One of your other episodes had a really strong health theme to it. I'm 65 years old so I found it a little funny, frankly, that Scott Bridges, the 34-year-old man who was the subject of this particular episode, was concerned that he was aging. But anyway, Scott began to take testosterone and human growth hormone in an effort to reverse the aging process. He got results that were predictable and noticeable, including growing muscles, but his physician reported that his liver was suffering. His sperm count dropped to nearly zero, which totally freaked out his wife. He ended up quitting. To you, what was the moral to that story?

MORGAN SPURLOCK: I think that the moral is that we live in a society where we're constantly being sold on a quick fix. Rather than work hard, rather than exercise, rather than change our diets, we look for a shot, a pill, or a surgery that will make us look better. For me the greatest lesson to take away from that show is that there are no shortcuts. We live in a society where we're so often sold the silver bullet every single day.

JOHN ROBBINS: Instant gratification shows up in highly sugared cereals that are right at kids' eye level at the supermarkets. It provides these jolts of flavor but no nutrition. The cells are being starved while the senses are being hyperstimulated.

MORGAN SPURLOCK: That is exactly what was happening to me while I was making *Super Size Me*. By two or three weeks into that diet of eating nothing but McDonald's food on a daily basis, I would eat a meal and like fifteen minutes later I would be starving again, because my body was craving real nutrition. When you get to the point where your body isn't getting what it wants, you'll overeat. And most of us overeat the very things that our body doesn't actually need to begin with.

So you're eating more and more because you keep saying, "I'm hungry so I'll have some more cookies, I'll have some more chips, I'll have another Coke." And the next thing you know, you're still hungry. I've never known anybody who says, "I'm going to eat more broccoli. I'm going to eat more bok choy or asparagus." Because once you're done with that, your body is like, "Oh great, I'm done. I'll let you know when I need some more."

JOHN ROBBINS: Well I have actually been known to overeat asparagus when it's in season.

MORGAN SPURLOCK: I will overeat anything when it's in season. In April, when it's strawberry season . . . there's nothing like fresh strawberries. I could sit and gorge myself on those all day long. I remember my grandmother had an apple tree in her backyard with those little small super sour Granny Smith apples. It was one of the greatest things. I was probably 10 years old, and the apples were just starting to fall from the tree. She'd say to me; "Don't you go eating all those apples." Sure enough, what do I do? I sit back there and I eat apple after apple until I get sick. Then my grandmother's like, "That serves you right for eating all those apples."

JOHN ROBBINS: Morgan Spurlock beginning his career. Before eating McDonald's for 30 days, he ate nothing but his Grandma's apples.

MORGAN SPURLOCK: The writing was already on the wall.

17

Nikki Henderson

Food Access in Historically Underinvested Communities

Nikki Henderson serves as Executive Director of People's Grocery, a nonprofit organization dedicated to providing fresh, good food to people of all income levels. She is also co-founder of Live Real, a national collaborative of food movement organizations committed to strengthening and expanding the youth food movement in the United States. In 2010 Nikki was featured in Elle *magazine as one of five "Gold awardees" for her innovative work in urban nutrition, agriculture, and enterprise.*

Nikki lives and works in West Oakland, a community of 30,000 residents that has fifty-three liquor stores and not one full-sized grocery store.

In the United States and many developed nations, consumption of highly processed sugar-laden "junk" foods, white flour, unhealthy fats, and artificial chemicals is highest in low-income communities. Not surprisingly, these same communities suffer from disproportionate rates of cancer, heart disease, diabetes, and many other health ailments. If you believe everyone should have health and nutrition, you'll be inspired to find out what Nikki and her community are doing to make healthy food accessible to everyone.

JOHN ROBBINS: You work is in West Oakland, which is a community that faces tremendous economic hardship. When people are struggling to survive, nutrition often takes a back seat to day-to-day survival. How do you respond to that?

NIKKI HENDERSON: Trying to think long term is a struggle for anyone whose basic needs aren't being met. Our strategy is based on listening first. The first time we interact with someone, we don't start off by trying to communicate our message. Instead, we focus on building the relationship. This is how we can find out what is important to people, and where we can connect. If I talk to someone long enough, I will be able to find out why healthy food is important to them. For example, my great aunt and great uncle are both amputees from diabetes, and that is what makes healthy food so important to me.

JOHN ROBBINS: An African American child in the United States today has a one in two chance of developing diabetes—significantly higher than the national average. Diabetes is deeply linked to obesity. It is also tied to a lack of food options. People's Grocery is based in West Oakland, which is a primarily African American community in which there are lots of places to get cigarettes, beer, and candy, but not a lot of options for people who want fresh vegetables.

NIKKI HENDERSON: Just dropping a grocery store randomly into a community won't necessarily help. People will probably buy the same things they got at the liquor store, just for lower prices. It is not like you magically want to go to the produce aisle if you haven't been eating produce before.

We need economic access, so that food is affordable. We need cultural access, so that the foods themselves are culturally appealing and people know how to prepare them and integrate them into their diets. We need physical access, so people have a place to get healthy food. And we need informational access, so people have the knowledge to help them make their choices.

People's Grocery brings a holistic approach to creating a food system. We make healthy food more affordable and available through The Grub Box (which is like an urban CSA). We also offer a Leadership Development Institute, where people can develop projects and campaigns with a foundation of cultural competency and deep relationship building. Some of our graduates have developed their own catering companies, cooking classes, or other projects in the community that raise the demand for healthy food.

JOHN ROBBINS: Are there programs that involve growing food where people are getting their hands in the dirt?

NIKKI HENDERSON: We have a couple of urban agriculture programs. One is at the California Hotel, which is a historic single-room occupancy hotel for low-income residents. It was once a site where African American artists could stay when they came to perform in San Francisco, back in the days of segregation. It has always been a place for people who didn't really have a place. We operate a garden in the backyard, with a greenhouse, chickens, raised beds, vermiculture, an aquaponic system, and a grey water system. Everyone from the community can come there to volunteer. We try to target it specifically to the people who live in and near the hotel.

We also operate the Growing Justice Institute, where we choose six to eight community members who have ideas for food and health projects. We support them for two years as they design and implement their project in the community. One of the projects that will launch soon is a community garden at a church in West Oakland.

JOHN ROBBINS: What is it that you love about the community that you work in?

NIKKI HENDERSON: I love how deeply saturated the history is in West Oakland. I can ask a random person on the street about the Black Panthers' free breakfast program, and they'll know where it was. The United Negro Improvement Association (UNIA) has an office here, and people are active and activated around developing their community. I walk to work on a regular basis, and on the way to work I will stop by the worker-owned Mandela Food Cooperative. I know all of the boys who work there. We'll laugh and they'll send me on my way with a kombucha.

Looking at the long-term history of West Oakland, the community in the early 1900s was mostly European immigrants. Then when the great migration happened from the South, the community became African American almost overnight. The community is fluid and accepting of new people, because it has always been that way.

One of the reasons we choose to be rooted locally is because the transformation of historically underinvested communities like West Oakland has to begin with the people who live here. At the same time, no local effort will work unless it is also tied to local efforts in other communities. So we are developing a statewide network of food justice organizations that can share best practices and learn from each other.

JOHN ROBBINS: What are some of the key things that you find effective when you are helping low income communities to learn about health and nutrition?

NIKKI HENDERSON: It is often most effective to start with asking questions, like: What do you know about healthy food? Do you eat healthy food? How often do you eat vegetables? How have you cooked vegetables in the past?

There is really no way to educate someone until you know where they are coming from and you can meet them where they are at.

I feel it is most effective to approach the community as a peer, and to build a partnership. Different people who live in the community will offer different things. Our Growing Justice Institute model is all about finding the ambassadors in the community who are passionate about healthy food. The program helps them to be spokespeople, and allows us to partner with them in transforming the community.

JOHN ROBBINS: You're working on a local level, but at the same time you're engaging in a national and even international conversation about systemic change. If you had the power to make large-scale changes in government food policy, what kinds of changes would you make, and why?

NIKKI HENDERSON: More and more people are depending on food stamps to feed their families. There is a big conversation about increasing the amount of money that goes to food stamps and other federally subsidized nutrition programs. My argument would be that if there is going to be an increase in federally subsidized nutrition program funding, it shouldn't all go to emergency services. It would be more effectively used in programs like ours that are trying to build self-sufficiency. We need to work towards a future in which people can get off food stamps, because there are local food systems that can support them. Then, as people get off food stamps and government funds are freed up, I'd like to see the funding that was going into emergency services utilized for another ten years to build up more robust local food systems. If you are funding emergency services, you need to also fund self-sufficiency programs so that people can eventually get off and stay off of emergency services.

JOHN ROBBINS: On a national and global level, we have food distribution networks that depend on very long supply chains. Fresh produce in the United States travels an average of 1,500 miles between farm and dinner plate. This system is made possible and economically viable by cheap oil. But oil is limited, and in the times to come, the cost of transportation will be increasing. In the future, these long supply chains may also be susceptible to rupture. When you eat locally you get fresher food, you save on transportation costs, your carbon footprint is reduced, and your food supply is more reliable.

The work of People's Grocery is a model, not just for low-income communities, but really for every kind of community that wants to source locally, to help people connect with each other, and to build connection around food and culture.

NIKKI HENDERSON: I believe there's a healthy balance between a local food system and regional, national, and international food systems. Trade can be a good thing. I think we are just imbalanced right now when it comes to the amount of regional and national entities in our food systems versus local entities.

JOHN ROBBINS: Compared to the national average, African American communities typically have much higher rates of high blood pressure, obesity, diabetes, heart disease, and many forms of cancers. Much of this is attributable to poor diet, and it is causing enormous suffering. Are there any particular kinds of foods that you like to promote?

NIKKI HENDERSON: We try to experiment and meet people starting where they're at. We find out what types of foods people like, and then show them ways of preparing those foods that will be helpful to whatever health ailments they are dealing with. We also incorporate a lot of salads, and invite people to try new things like mixing fruits and vegetables in ways that they might never have thought about.

JOHN ROBBINS: You work in the presence of some painful realities. In West Oakland, the suicide and homicide rates are high. Life expectancy is low. People can start to feel that life is cheap, and that there really isn't much hope. How do you keep yourself lit? What keeps the song in your heart alive?

NIKKI HENDERSON: I have a pretty intense spiritual practice that involves Native ceremony and lots of meditation. My prayer practice forces me outside of my box, and provides a deep place to resource from when confronting trauma and crisis. I also receive a lot of support from my community, my fiancé, and my family. I call my grandmother on the phone and pray with her twice a week.

At People's Grocery we focus on community celebration, because there is always something to celebrate and there is always something to grieve. For healing to happen, we need to devote intentional time for both. We hold a lot of dinners. Sometimes there are a lot of tears, and other times they are filled with laughter. Everyone knows that they can come and be loved and supported.

JOHN ROBBINS: It's so true. There is always something to celebrate, and there is always something to grieve. Sharing our joy and our sorrow can connect us to each other in an authentic and real way. It also helps to awaken our power to respond to the problems and celebrate the beauty in a way that brings more life to us all.

Steps You Can Take: Action for Systemic Change

Sign or Start Petitions

Check out and sign existing petitions sponsored by organizations like *www.care2.com, www.change.org,* or *www.signon.org.* All three of these organizations also make it easy for you to launch your own petition. The petition organizations generally deliver messages to targets via email. If you get a lot of signatures, you can deliver them in person, and contact the media to bring attention to your issue.

Lobby Your Representatives

Call the office of your political representatives and ask if you can set up a constituent visit. Most likely, they will book you a slot with them or a member of their staff. You can use the meeting to make your case and ask for their support.

Join Others

Join organizations working legislatively, and participate in their campaigns. Many food revolutionary organizations are listed at *www.foodrevolution.org.*

Vote with Your Dollars

Participate in boycotts. Take your money away from causes you oppose—and write the company, or post on their Facebook page, telling them why. Redirect your money towards "buycotts," by purchasing products made in ways that are more consistent with your values.

Write Letters

Write letters to companies, telling them what you think of their practices. Complain to companies that bother you, but also celebrate companies that are doing good things.

Spread the Word

Spread the word amongst friends and colleagues. Give them resources, books, videos, and tools. Empower them with as much information as they're ready to absorb.

Make Good Food Fun for Kids

Research suggests eating habits are developed early on in a child's life—oftentimes before the age of 6. Super Sprowtz bridges the gap for parents and educators who are looking for tools and ways to engage their families and students to eat more vegetables and make healthier food choices. Their motto is Good Food Made Fun. Check out their amazing library of videos, curricula, and other resources by visiting *www.supersprowtz.com*.

Take Action in Your School District

You can get excellent tools and resources to help inspire your school district to take healthy steps from two great organizations that have websites at *www.healthylunches .org* and *www.HealthySchoolLunches.org*.

Resources for Transforming Food Policy

RajPatel.org

www.rajpatel.org

Dr. Raj Patel is an award-winning writer, activist, and academic with expertise in global food policy and the relationship between food and justice. Raj's website hosts a blog written by Raj and updated regularly. There are many current and archived videos and downloadable articles of Raj presenting on the global food crisis and other topics from his books. He also provides a list of helpful links and a reading list of books that he recommends.

Appetite for Profit

www.appetiteforprofit.com

Michele Simon is a public health lawyer who has been researching and writing about the food industry and food politics since 1996. She specializes in legal strategies to counter corporate tactics that harm the public's health. Michele offers a blog about current public health topics and she exposes truths that food corporations try to keep hidden. The website provides a collection of articles, videos, and newsletters.

Kucinich Action

www.kucinich.us

Kucinich Action was founded by Congressman Dennis Kucinich with the goal of empowering individuals to engage with the political process. The organization endeavors to build a powerful grassroots movement to return the power to the people in America's political system. Kucinich Action advocates for dynamic initiatives and identifies and supports bold, independent-minded leaders.

GMO Film Project

www.gmofilm.com

Led by Elizabeth Kucinich and a team of partners, this film project (still in progress at the time of this book's publication) tells the story of poor Haitian farmers burning seeds in defiance of Monsanto's gift of 475 tons of hybrid corn and vegetable seeds to Haiti shortly after the devastating earthquake of January 2010. The story goes on to ask why hungry farmers would burn seeds, and the insights gleaned shed light on what has happened to food in the United States, and why the GMO struggle has profound implications to the entire global food supply.

People's Grocery

www.peoplesgrocery.org

Directed by Nikki Henderson, People's Grocery is a leader in the movement for food justice. Their mission is to improve the health and economy of West Oakland through the local food system—and they are a model for many other organizations. People's Grocery's urban agriculture, nutrition, and enterprise programs provide healthy food access while setting the stage for a systemic conversation about healthy food for everybody.

The Coalition of Immokalee Workers

www.ciw-online.org

The Coalition of Immokalee Workers (CIW) is a community-based organization of mainly Latino, Mayan Indian, and Haitian immigrants working in low-wage jobs throughout the state of Florida. They work for, among other things: a fair wage for the work they do, better housing, stronger safety laws, stronger enforcement of workers' rights, the right to organize on their jobs without fear of retaliation, and an end to involuntary servitude in the fields. They run the Fair Food Program and the Anti-Slavery Campaign.

Growing Power

www.growingpower.org

Growing Power transforms communities by supporting people from diverse backgrounds and the environments in which they live through the development of Community Food Systems. These systems provide high-quality, safe, healthy, affordable food for all residents in the community. Their goal is a simple one: to grow food, to grow minds, and to grow community.

PART VI

Food and the Human Spirit

O ur food is often treated as a commodity more than as a sacred gift of the living Earth. It's hard to overstate the price farmers, farm workers, farm animals, and everyone who eats food are paying for this disconnection.

Whether we realize it or not, every time we eat, we are participating in a web of relationships that extends around the world. We are interacting with farmers, with the rainforests, with the air and water—and with ourselves. Bringing consciousness to this intimate interaction with our world and our own biology can bring healing and richness to our lives.

When you decide that you want to do something positive, that choice spreads out and makes you feel like a more powerful being. Bringing consciousness to your relationship with food might change more than your grocery list. Consequences could include greater health, unexplainable joy, a sense of palpable fulfillment, and even more loving relationships.

18

Frances Moore Lappé
Choosing Courage

*Frances Moore Lappé was one of the first people to popular-
ize the link between personal food choices and global hunger.
She is the author of eighteen books, including the three-
million-copy culture-changer* Diet for a Small Planet, *and*
EcoMind: Changing the Way We Think, to Create the World
We Want. *She is co-founder of Food First: The Institute
for Food and Development Policy, and of the Small Planet
Institute, which is a collaborative network for research and
education that is helping to bring democracy to life. A world-
renowned spokesperson for food justice and civic participation, Frances was named by* Gourmet *magazine as one of
twenty-five people who have changed the way America eats.*

*Frances believes that hope is not what we find in evi-
dence, it's what we become in action. In a world all too
often beset by apathy and resignation, what has she learned
about the roots of real democracy? Here are her insights to
help you choose courage, and to live with purpose.*

JOHN ROBBINS: Have things gotten better or worse since
you wrote *Diet for a Small Planet?*

FRANCES MOORE LAPPÉ: Both. It's a question of where we put our focus. I believe our focus is all-important, especially now that neuroscientists are telling us that our thoughts really do shape even the physical structure of our brains. There are as many hungry people in the world as there were when I wrote *Diet for a Small Planet*. Yet we know vastly more today, and we have proof of what works when a truly democratic food system is aligned with nature. So I try to grow my heart big enough to hold it all, not denying the negative, but keeping my focus on what is possible in the future.

JOHN ROBBINS: You try not to avert your gaze, to be really open to the pain and the loss that's taken place while still staying connected to what's possible.

FRANCES MOORE LAPPÉ: Yes, I say that I'm not a pessimist and I am not an optimist; I am a possibilist.

I think that's a really important distinction, because I can't predict the future. But I do know that the life force is so strong—that life loves life. From an ecological worldview—grounded in the reality of connection and continuous change—it's not possible to know what's possible. In fact, change is the only constant. So we have to believe in the possibilities.

JOHN ROBBINS: A recent report from Save the Children tells us that there are now nearly a half-billion children worldwide who are at risk of irreversible damage from malnutrition, stunted growth, and undeveloped brains. They say that chronic childhood malnutrition kills more than 300 children every hour of every day and affects one in four children globally. For decades you have been pointing out that there is enough to go around; that we suffer not from a shortage of food, but from a shortage of justice and democracy. There is really enough food, but some of us are consuming massive amounts of grain by cycling it through the animals we eat, while the world's hungry don't have access to food supplies because they can't afford to buy them.

In writing *Diet for a Small Planet,* one of the great contributions you made to our common understanding was to point out that modern meat production—particularly feed-lot beef—has become a protein factory in reverse. What can we do to become part of the solution to this problem?

FRANCES MOORE LAPPÉ: First we need to shed this very dangerous myth that seems to have its grip on so many people—the idea that there is just not enough. We're taught to see scarcity everywhere. But once we let go of the false belief in a quantitative lack of food, then we get curious and ask, what are the root causes of hunger throughout the world? Curiosity is a powerful antidote to fear.

One of my favorite words in the English language now is the word "align." At present, our economic laws and our assumptions about our nature are perversely aligned with what is true about our nature and what works in nature. The dominant corporate-controlled, chemical-dependent agricultural system is failing. It is failing both in the "wealthy" countries and in the poor countries. The answer, then, is to begin to align our own daily choices with what is best for the earth, best for our bodies, and best for other human beings.

And what's so great is that with food, they are all the same thing.

Part of the theme song of my life is also about rethinking power. Power in its Latin root simply means "capacity to act." When I'm facilitating a workshop and I ask people, "What are your associations with power?" People's answers are usually really negative. I think part of the food revolution is reclaiming power as a positive: a deep human need.

JOHN ROBBINS: When we claim our power rather than fearing it, we grow in our capacity to create, and also in our ability to prevent and alleviate suffering. One of the things that your work has highlighted for me is how the same food choices that are best for the environment—those choices that allow the most food resources to be available so that more people can eat—also tend to be the healthiest and the most earth friendly.

Eating locally is a good example. Locally grown food is fresher and more nutritious, it tastes better, and the carbon footprint is lower because less transportation is required. Eating more organically also reduces our carbon footprint. In addition to being toxic to the soil and to ourselves, most of our synthetic fertilizers and pesticides are made from oil and natural gas. By eating lower on the food chain—getting more of our protein from plants and less from animals—we greatly decrease the greenhouse gases that are emitted in the production of our food, and at the same time reduce our rates of heart disease, obesity, and many other health problems. It's a major win-win. There are so many positive things here.

FRANCES MOORE LAPPÉ: That was such a thrilling discovery when I was working on *Diet for a Small Planet* in the 1970s. As soon as I started having the sense that a plant-based diet was part of the answer, I thought, "Well of course, if I just share this information with people, everybody will say, 'Yes!'" But it wasn't that easy. Back in the 1970s, eating a plant-centered diet was considered really heretical, especially for someone like me, who grew up in cowtown USA—Fort Worth, Texas. The idea that we could thrive without meat was really a crazy idea.

JOHN ROBBINS: You have said that we face a crisis of democracy. Why aren't our political leaders standing up for the well-being of their constituents?

FRANCES MOORE LAPPÉ: There are roughly two-dozen lobbyists for each person Americans elect to represent us in

Washington. Elected officials tend to pass laws that mirror what the lobbyists want. The lobbyists, for the most part, represent the interests that benefit most from the status quo. In this way they continue to reinforce the concentration of wealth, to pull back environmental rules that could protect us, and to stand in the way of enforceable standards pertaining to the quality of our food and the advertising of junk food to children. I see it as a cycle: the narrowing of information and the increasing impoverishment of people makes it harder and harder to purchase healthy foods.

JOHN ROBBINS: In *Hope's Edge*, you and your daughter Anna write: "Hope doesn't come from calculating whether the good news is winning out over the bad; it's simply a choice to take action."

FRANCES MOORE LAPPÉ: Yes, and we simplified that as the slogan on our website: "Hope is not what we find in evidence; it's what we become in action."

This was our great takeaway from a global journey Anna and I took together. Whether we were working with Vandana Shiva, or hiking up in the foothills of the Himalayas, or with the late Wangari Maathai in the villages in Kenya, one realization changed our lives. It was that the most hope-filled people, the people with the most energy and verve, were not the people with the least problems. In fact, they were often people with great challenges. But they were moving; they were in action with others. They were doing everything they could and beautifully celebrating what they were accomplishing.

If we try to seek out others who are in action, those who are already engaged, and believe that we will become more like them as we move shoulder-to-shoulder together, that is how real hope, or what we call honest hope, emerges. Again, it's not just looking for proof, it's actually creating it.

JOHN ROBBINS: One of the ways that we can break out of resignation and passivity and find the joy and the vitality that comes with taking action is to seek out people who are already engaged in action in a vital way, and join them and be part of what is already occurring that's positive and productive.

FRANCES MOORE LAPPÉ: We are organisms, we human beings, and like every organism we're in an ecosystem. So much of any organism's behavior, drive, or lack of drive is dependent upon the stimuli. So we are shaped by what we bring into our lives. As we observe others taking risks, as we move outside of our comfort zone, try new things, and bring new people and new stories into our lives, we are changed.

If we want to become more courageous we've got to hang out with courage.

It's so exciting to think that we are in large measure the product of the stimuli in which we live. That means we have choice. We can consciously choose what's shaping us. When I say that, the person who comes to mind is my hero, Wangari Maathai, who died in 2011. When Anna and I met her in Kenya, she changed our lives.

In the 1970s Wangari founded the Green Belt Movement, which organizes women in villages to plant trees—now more than 50 million trees! Sometimes the women had to stand up against their tribal chief and sometimes against their husbands to really make this movement work. They all wore a simple white T-shirt that said, "As for me, I've made a choice."

Anna and I always found that so powerful. You can complain and you can feel down. But at some point you can make a choice—a choice to act. It may be something small or large, but once you've made a choice, that will lead to other choices.

JOHN ROBBINS: As for me, I've made a choice.

FRANCES MOORE LAPPÉ: As for me, I've made a choice.

19

***

Kathy Freston

Leaning into a Healthier Life, One Bite at a Time

***

Kathy Freston is a bestselling author with a focus on healthy living and conscious eating. Her book, Veganist: Lose Weight, Get Healthy, Change the World, *was an instant* New York Times *bestseller, as were three of her other books—*The Lean, Quantum Wellness, *and* The One. *A frequent guest on television shows like* Oprah, Ellen, Dr. Oz, Good Morning America, *and* The View, *Kathy promotes a body/mind/spirit approach to health and happiness that includes a concentration on healthy diet, emotional introspection, spiritual practice, and loving relationships.*

Kathy invites you to eat consciously as a way to feel empowered and on track with your personal growth. She asks: How can we say we're striving to evolve spiritually, without considering how the food we're consuming got to be on our plates?

JOHN ROBBINS: Do you see a relationship between our diets and our spiritual integrity?

KATHY FRESTON: Absolutely. For me, my brand of spirituality is pretty straightforward; it's just a matter of living as true as I can to the basic principles that run throughout all the wisdom traditions: kindness, mercy, alleviating of suffering, and responsible stewardship. In every area of my life, I ask myself what sits right in my soul, and then try to make choices aligned with that.

In my earlier work I encouraged people to look closely at their relationships to see how they could use them as vehicles to evolve personally. I wrote about meditation and spiritual practice, introspection and conscious relating; but I realized I wasn't addressing the thing we all do every day, several times a day: eating. If one is interested in evolving as a human being, how food gets to the dinner plate has to be considered. How can we leave such a big thing out?! If we are striving to be ever more aware and awake, we can't stay ignorant about such a big part of our lives. And food *is* a big part of our lives! To be spiritual, after all, is to be conscious—of your actions, of the effect you're having, and of what others go through.

Now, growing up in Georgia, I ate chicken-fried steak, burgers, barbequed ribs, cheesy grits, and vanilla milkshakes. I was part of a good family and nobody was bad just because we ate this way. We simply didn't think about it. I carried that obliviousness into my adulthood. When I finally realized that I needed to look at the impact of my food choices, I thought: "This is really uncomfortable. It hurts to see and read about this stuff." Bearing witness was the hardest awakening I've ever experienced, but also the most profoundly moving one. Because I thought, "Wow, I can actually do something about this. I can stop eating animal products. I can stop supporting an industry that causes suffering." And so even though it was uncomfortable, it was hugely empowering to see the truth and then decide what felt right in my soul. It was a process that brought me deeper into my spirituality, deeper into my personal evolution.

When I started reading books like *Diet for a New America,* things became so clear. It was a real awakening for me to see the extent of damage that the business of eating animals caused—both to human and global well-being. Your book really brought everything together for me, and I realized I had to put my money where my mouth was and begin moving away from animal food. After all, if my spiritual goal was to become a more caring and concerned person, how could I continue to knowingly cause pain?

JOHN ROBBINS: There are a lot of people who try to live by values of kindness and compassion, and who want to alleviate suffering when they can, but who would rather not look at what is done to most of the animals that are raised for meat and dairy products. Of course it is all done behind closed doors, and the industry would just as soon people not look. What do you see when you lift the veil? And do you see a connection between what is being done to the animals —we call them livestock—who are being raised for food, and something happening to our spirits?

KATHY FRESTON: Most of us are empathic and compassionate by nature, I think. So if we're eating meat and dairy, we have shut down our awareness. We have to deaden ourselves. Being deadened is the opposite of being enlightened. I don't want to be dulled to the truth; I'd rather be open and aware and a force for healing. When we let ourselves be affected by what we see, something comes alive in us. It's like we step into our higher nature; like we vault ourselves up to a new sense of empowerment. Whenever I've challenged myself to rise to some occasion that I absolutely didn't want to rise to, I've found that life seems to meet me more than halfway. Whenever I've pushed myself past my comfort zone, I've always felt that there was some unexpected reward or good feeling at the other side of it. Like life was just waiting for me to make a bold step and move away from old habits that no longer made sense. I sometimes feel like the whole universe is conspiring to help push us forward, and if we just take the first steps ourselves, this wonderful energy comes rushing into our lives. In this case, those first steps were to lean away from animal foods and toward plant-based foods.

JOHN ROBBINS: We sometimes get stuck in lethargy or paralysis, and resign ourselves to that. But once we engage, mobilize, and start moving, then there is a momentum that takes over and we get swept into larger streams of energy. My experience has been that when we stop playing small, more is asked of us. We get bigger, and we become more accountable.

KATHY FRESTON: Agreed. I believe that eating a plant-based diet can help us become better aligned with our higher nature. When we eat food that's grown in the ground or on trees, we are eating peacefully. Not causing any suffering, not killing. We're putting into practice the things we'd like to see more of: peace, kindness, smart stewardship. Eating is so foundational; it's so elemental. So when we eat consciously, we're putting into regular practice the principles—spiritual principles, if you will—that we believe in.

But when I say "eat plants," I'm not talking about just boring ole green stuff. I grew up not liking spinach or broccoli. Sometimes it's a blocking point for people who think, "God, do I have to just sit there and eat a pile of green food?" But there are so many wonderful foods you can eat. You can have black bean burritos and meatless meat chili. You can have delicious pizza with nondairy cheese and veggie sausage, and you can have all kinds of burgers and hot dogs—just better versions of the things that you may have grown up with.

Here's the thing: we have this evolutionary impulse inside of us to grow, and whether or not you set out to do it, there is something within you, like a little homing device, that keeps pulling you forward. When you answer that call, something in you resonates, and your whole being lights up. You get this sense of esteem and confidence, and you feel that you are rising to something.

JOHN ROBBINS: You have been on *Oprah* several times, and she has been very moved by the way you frame your food choices in a spiritual context. At one point you helped her through a 21-day vegan cleanse, and afterwards she wrote: "I learned a lot about how animals are treated and mistreated before they get to our table. It is appalling and beneath our humanity to allow the torture of animals for the sake of our gluttony. We have neglected basic human decency on such a large scale, and it really does bleed over into every other aspect of life." What do you think Oprah meant by: "It really does bleed over into every other aspect of life"?

KATHY FRESTON: When we open our hearts, that translates into every aspect of our lives. We feel more compassion and more empathy. It makes us feel like we are more powerful beings. It makes us feel like we are kinder human beings. We are smarter, we are more aware, and we are hipper. We step onto the forward edge of human evolution. That is going to translate to being better partners in our relationships, and to attracting better people into our lives, because we are vibrating at a higher frequency. It is going to make us feel more directed in work, because we are going to feel inspired to push towards greater things. We are going to be energized, and we are going to be able to work harder.

Who we are as human beings shows up in every arena of our lives, whether it is our work, our relationships, our health, or our food choices.

JOHN ROBBINS: What do you see as the biggest obstacles that most people face in bringing their food choices into alignment with their spirits?

KATHY FRESTON: The first obstacle is one of awareness. Financial interests try to obscure the data, just like cigarette companies used to try to confuse us about the health impacts of smoking. But thanks to the great undercover work by some of the animal organizations, more and more people can see for themselves what's really going on in factory farms. And now there's a vast volume of science that is coming out in support of a plant-based diet, demonstrating clearly that it can be much healthier and that it can help prevent and reverse some of the chronic diseases of our time.

The other major obstacle is the availability of healthy food. I think the market is a little behind on that. Most of the time I am traveling, and I go to a lot of mainstream restaurants. I often find that 99 percent of the menu is meat- and dairy-oriented. I think that when mainstream restaurants and chains start making more delicious plant-based, nonanimal foods available, people will choose them.

When awareness meets availability, that is when things will really change.

JOHN ROBBINS: Many people think of dairy products as relatively cruelty-free. But in *Veganist* you make the point that animal abuse takes place on a massive scale in large dairy farms.

KATHY FRESTON: Dairy was the last animal product that I gave up, because for a long time, I thought that no animal was being killed for milk. Then I had a friend who told me that the male calves that aren't used in the dairy industry are sold off for veal. I was like, "What?! You have ruined my cheese now!" But I started Googling, and I saw what happens to these poor dairy cows. They are hooked up to machines, and they are tortured by having their little baby calves taken away at birth. Then they are kept in these tiny places, and they get infections in their udders. When they are used up and their milk production goes down, they get dragged off to be slaughtered for hamburger meat. So they are in fact killed, just like beef cattle—they are just tortured a little bit longer before their flesh reaches our plates.

Dairy for me was addictive, because I liked the comfort, the gooeyness, and the creaminess. There was something about it that made me feel cozy and warm.

But luckily, we now have these fantastic nondairy cheeses available that did not exist even a decade ago. My favorite is Daiya because it melts really well, and I love pizza. I get a pizza crust from the health food store, and I put on some pizza sauce and tomato sauce. I sprinkle some Daiya nondairy cheese on top, and then I chop up some veggie sausage and maybe some mushrooms and I pop it in the oven, and voila! I have my pizza, and no one would know the difference.

It is all about, for me, just leaning forward. When I gave up animals I didn't do it all at once. I gave up one animal at a time, and instead of substituting another animal product, I substituted a plant-based product.

Unless you are a really good chef, chances are you're like me and you have maybe seven to ten things that you just cook over and over again. We are busy, and we don't have time to be that creative. So I would just take one thing each week, and find an alternative.

For example, I used to have chicken with mashed potatoes and green beans for the family dinner every week or so. Now instead of having chicken, I started having one of those wonderful meat alternatives like Gardein, and then I would make the mashed potatoes with Earth Balance. Earth Balance is not margarine—it's made from cold pressed oils. But it's nondairy and it tastes exactly like butter. I would make the mashed potatoes with a little soy milk or almond milk. Then I would have my dinner that looked pretty much like the thing that I grew up loving, but it was just a better version. It was more conscious, and it was healthier too—with no cholesterol, very little fat, and lots of protein. So I got my comfort, I got my ease, and I avoided the thing that was doing so much damage.

JOHN ROBBINS: I remember when I wrote *Diet for a New America* and it first came out. I got a phone call from a very fine spiritual mentor of my generation, named Ram Dass. He called me and told me that he had read my book. Then he said: "You have ruined my dinner! I can't eat chicken anymore. How dare you?!"

KATHY FRESTON: It's true, that's the ultimate inconvenient truth, isn't it?

JOHN ROBBINS: For me, the hardest thing to give up was ice cream. I guess that is understandable considering that I grew up in a home with 31 flavors of ice cream in the freezer. What helped me finally to no longer hanker after it was finding some wonderful, nondairy frozen desserts. There are a lot of them now, and it made it easier for me. Being a work in progress myself, and with all the weaknesses and flaws that are part of the human condition, I really appreciate having those alternatives.

KATHY FRESTON: I think perfection is the enemy of the good. When we hold ourselves and people around us to strict rules, we set ourselves up for failure. The more that we can allow ourselves to just keep moving forward and live by, "Progress, not perfection," the easier it will be. I know a lot of people are very against the so-called faux foods, saying that they are processed or that they are not whole and completely healthy. But I've got to tell you, if we didn't have those foods, many of my friends and family would have never gone down this road.

If you don't have alternatives to crowd out those old favorites, then for the most part, unless you are highly motivated, you are not going to be able to do it. The taste buds, cravings, and comfort culture win out. So when you say: "I am going to have that veggie sausage," or "I am going to have that faux chicken," or "I am going to have that nondairy cheese," it might not be the ideal of brown rice, beans, and vegetables, but it is a great step.

And then maybe a few months down the line, you might find yourself saying: "I am feeling a little bit cleaner. I want to eat a little bit more whole foods, so I am going to work in a few dishes that are really wholesome." And that is how it happens. It is a lifelong continuum of just pushing yourself ever so gently forward.

JOHN ROBBINS: As human beings, with all our inevitable imperfections, we crave compassion. People will sometimes say to me that they have made a mistake and I say, "Good, just let the mistake lead you, eventually, in the direction of your intention."

KATHY FRESTON: When I started this process, I was a girl from the South who ate everything in sight without thinking about it. Then I decided I didn't want to eat anything from an animal, and I was completely confounded as to what to do. So I said, "You know what? I am just going to set my intention, and my intention is to be someone who doesn't eat animals. I don't know how I am going to get there, but I am going to nudge myself forward and lean into it." That intention continues to push me through, because I am still not perfect. I still do things that I think are unconscious. But life is long, and I continue to keep growing.

Back in the day when I was introduced to this way of thinking, if someone came up to me and said: "You shouldn't do that," and told me I should feel bad for it, I would have shut down. We tend to either like the messenger and be open to their message, or shut down to them and reject their message. As ambassadors of a food awakening, it is beholden on us to be kind, warm, patient, and generous with our spirit and with our information. And remember, we—all of us—are always works in progress.

20

Marianne Williamson
Food, Body, and Divine Perfection

Marianne Williamson is one of the world's most beloved spiritual authors and lecturers, and the author of six New York Times *bestsellers. She is founder of Project Angel Food, a meals-on-wheels program that every day serves more than 1,000 homebound people with AIDS. In her latest bestseller,* A Course in Weight Loss: 21 Spiritual Lessons for Surrendering Your Weight Forever, *Marianne brings to food and weight loss the same piercing insight and spiritual principles she brings to so many areas of our lives.*

What are Marianne's top insights to help you bring love and consciousness to your body, your relationship with food, and your own healing journey?

JOHN ROBBINS: You are using principles that have been established and manifest strongly in recovery and twelve-step movements, and applying them to weight loss. How does that work?

MARIANNE WILLIAMSON: When you are dealing with a serious compulsion or addictive pattern, then by definition self-will, self-discipline, and any other machinations of the conscious mind are not enough by themselves to handle the problem. It is like a breaker switch in your brain is simply flipped. Anybody who has had this kind of a problem knows that it doesn't matter how intelligent you are. Sigmund Freud said, "Intelligence will be used in the service of the neurosis."

Often people will think, "I know enough not to do this stupid self-sabotaging thing, so why do I keep doing it?" When you look at obesity in the United States, clearly it is not a bunch of stupid people. It has nothing to do with intelligence. Sometimes people who are dealing with issues of obesity and compulsive eating know more than I will ever know about nutrition, metabolism, and exercise, because they have studied it. But clearly the real problem, and therefore the real solution, is on another level of consciousness, and that is where the spiritual work comes in.

JOHN ROBBINS: In *A Course in Weight Loss*, you wrote that "This course is not about your relationship with food, it is about your relationship with love."

MARIANNE WILLIAMSON: Your relationship with love is your relationship with the essence of who you are. It affects your relationship with your body, and your relationship with food. When you realize that you are a spirit and that this body is a temple, then you want to treat it well. Once you see that everything in life is a gift, you see that food is also a gift. If you are about to eat chemically processed, unhealthy food, you realize that to do so is, on a certain level, an act of violence against yourself.

JOHN ROBBINS: Do you think that as people increase their sense of connection to their source, to their true nature, and to their body, that they will then develop a more healthy relationship with food?

MARIANNE WILLIAMSON: Oh, absolutely! People who are meditating every day and involved in a serious spiritual practice don't usually wake up in the morning and want to rush out to eat a bunch of junk food.

JOHN ROBBINS: I have seen in some spiritual circles, the idea that what you think determines pretty much everything, and so to concern yourself with things like nutrition and eating healthfully is to embody a lower level of consciousness.

MARIANNE WILLIAMSON: If I am eating something that is unhealthy for my body and irreverent towards life, that is a thought. And all thought creates form.

JOHN ROBBINS: What do you see as the spiritual underpinnings of widespread hunger and malnutrition?

MARIANNE WILLIAMSON: Mahatma Gandhi said, "The problem of the world is that humanity is not in its right mind." The cause of so much that we struggle with today stems from a fundamental separation from love. There are 17,000 children on this planet who starve to death every day. They are part of the hungry bottom billion on the planet, the people who live on $1.25 and less a day. Now the economist Jeffrey Sachs has statistically proven that for $100 billion, spent over a ten-year period of time, we could eradicate deep poverty from the planet. So here we have $700 billion a year that we spend on defense in the United States, and for one-seventh of that, spent over ten years, we could eradicate deep poverty.

It all goes back to a fundamental insanity and lack of love. We have so many people who are undernourished or malnourished or starving, and that fact is juxtaposed with the fact that at least as of now, we have enough food.

JOHN ROBBINS: Scientists fear that global warming could create 150 million environmental refugees by 2050. And while, despite distribution problems, we do grow enough food to feed humanity now, climate disruptions and a rapidly growing human population may change that in the not-too-distant future. To look at our collective predicament through the lens of a twelve-step program, I sometimes think that the modern industrialized world is like an individual who is addicted or who is possessed by their alcoholism, and is very near hitting bottom. But we haven't yet turned over our consciousness to what is greater and higher and truly possible—to the power of love. Following that metaphor, do you see that we are capable of collectively becoming sober?

MARIANNE WILLIAMSON: There is always a possibility, and I think that hope is a moral imperative. Do I see humanity bottoming out? Do I see humanity having a mass experience of all of us looking at each other and saying, "Wow, let's do it another way now?" Absolutely—it's inevitable that it will happen. The issue is how much human suffering will have to occur first. I mean if we start lobbing nuclear bombs, and eventually there are only fifty people left in the world, they will look at each other and choose to go on another way. How much human suffering has to occur first? I think that is up to us.

JOHN ROBBINS: There are a couple of bumper stickers that I have seen. There is one, "If you aren't outraged you aren't paying attention." And then there is another one, "If you aren't in awe, you aren't paying attention." In some way, it seems that our hearts are being stretched to span the violence and the suffering in our lives and world, and at the same time we are also witness to so much magnificence and brilliance. We are a species that has produced nuclear weapons, and also one that has developed profound insight into the nature of life itself.

MARIANNE WILLIAMSON: I sometimes say that if you are not grieving, you are not conscious. But if you are not rejoicing in the possibilities of how this could all change, then you are not looking through the filter of the greatest spiritual perspicacity. Both grief and awe are realistic feelings about the state of the world today. The issue now is about what we are going to do. In every moment, we make a choice about which direction we are going to go. Once we realize that this is a self-correcting and self-organizing universe, then we don't have to look any further than our own life circumstances, the relationships and situations that we are in, to begin the transformation.

JOHN ROBBINS: There is a belief behind our disrespect for the earth, for our feelings, and for ourselves as members of the earth community. It is a belief that we are here to dominate, we are here to control, and that the earth and the living web of life is subservient to us.

MARIANNE WILLIAMSON: The beginning of the environmental crisis in Western civilization really began with the early church's systematic destruction of pagan culture. Because in pagan culture, part of the role of women as the High Priestesses (or witches, as they were sometimes called at the time) was to hold a sense of divine and sacred partnership with the earth. One of the reasons the effort was made to systematically destroy that culture, was because the early church was seeking to introduce the dispensation that we were in fact not in divine partnership with nature, but rather that God had given it to man to dominate nature.

That is when a very insidious and poisonous thought form took such strong hold in Western consciousness. Yet today even ardent adherents to the dominator thought will often agree that we need to be good stewards, and I think that that is really where the conversation needs to begin.

Some of us believe humanity should be in divine partnership with nature, some people believe that man has been given by God the right to have dominion over nature. But since even they say that we should be good stewards, that right there should be the common ground.

Recently I was at a conference sponsored by The Humane Society. It was about abuses to animals and laws that need to be changed. It was a rather small group, and it included a very high-level group of people who were like the head of the Southern Baptist Convention and other serious players in conservative Christianity. What I noticed there, and it was absolutely fascinating to me, was that if you say to those people that the same spirit of God is in an animal that is in you and me, they bristle and shut down. However they do believe that God has instructed us to be merciful, kind, and tender towards the animals. Within that context, I was amazed and deeply respectful of how politically organized and active they are regarding laws that will regulate the treatment of animals.

We were watching these videos of horrible animal abuses. I saw a woman who heads the Moral Majority or some similar organization, sitting there with tears streaming down her cheeks.

JOHN ROBBINS: I often think about the biblical injunction that we are given dominion over the animals. What does that mean? If I have two sons and I go out for the evening and I say to the older one, "You are in charge while I am gone," and I say to the younger one, "Will you please do what your older brother says until I get back?" I am in effect giving the older boy dominion over the younger one for the time being. But I wouldn't be happy if when I returned home I found that the older son had tortured, cooked, and eaten the younger one; or performed some macabre medical experiment on him; or had been otherwise less than merciful, kind, and tender towards him. Even if we accept dominion, what does it mean to be reverent and compassionate?

MARIANNE WILLIAMSON: There is a responsibility that comes with dominion. If we are supposed to be good stewards of the earth, then we should ask: How is it good stewardship to destroy the Amazon? How are factory farms good stewardship?

JOHN ROBBINS: And how are we being good stewards of our bodies when we eat poorly and abuse them? Yet, many of us carry compulsions that arise from levels of the psyche and from our relationship to life that are beyond self-will. Some of these compulsions stem from childhood. How can we help our children grow up with good food habits, confidence in themselves, and a positive body image?

MARIANNE WILLIAMSON: All of us as parents see on a daily basis the way that our own choices affect our children. My daughter is 21 and at about age 20 the jury comes in. At that point, I felt like I could see where I got it wrong and where I got it right. I look back now and think that in parenting, you have to know when to hold and when to fold. One of the places where I folded and wish I hadn't was with food. I remember that when my daughter was an adolescent, she used to bring all these chips and empty carbs into the house. I would say: "We don't have things like that here. We have fruits and nuts and vegetables." She would reply: "But all the other kids do and I have to have them when my friends come after school or I'll be a total dork." That is a place where I folded and I wished I hadn't.

We are responsible for what we have in our homes, and what we serve our children. But I do work with a lot of women who say: "I want to have healthy food in the house, and I want to feed my children healthy food. But my husband insists on junk." It can be hard for a lot of people, even when they have the best of intentions, to fill their homes with healthy food.

JOHN ROBBINS: There is an intricate dance between will and surrender—between determination and unconditional love. This shows up in families, and also in relation to our own bodies.

The fat acceptance movement holds that people are unfairly discriminated against because of their size and appearance, and that this discrimination is a form of violence. It seeks to change discrimination in employment and cultural stereotypes, and also to help fat people come to greater levels of self-acceptance.

I want people to feel good about themselves. There is beauty in everyone, of every size and shape. There is beauty in people at every level of health and illness. At the same time, I want people to be healthy. I know that weight is not just a question of vanity or cultural conditioning. There is a tension that I think many people feel between self-acceptance and unconditional love on the one hand, and the courage to change what we can on the other hand.

MARIANNE WILLIAMSON: It is a basic truth that a situation has to be accepted fully before you can have the power to change it, so accepting yourself where you are is important. However, there are real health implications to obesity. I often tell people: "You can hear it from me now, or you can hear it from your doctor later." There is no way that carrying fifty or sixty extra pounds is easy on your heart, your lungs, or your liver. That's a fact. Every person in the world, no matter what size, shape, or form they are, deserves respect and love. But that doesn't mean we are supposed to pretend that something is healthy when in fact it is not.

JOHN ROBBINS: What do you see as the relationship between the body and the spirit?

MARIANNE WILLIAMSON: How we sit within the body is an extremely important part of the spiritual journey. The body itself is used either by the spirit within us, or by the fear-based mind. When it is used by the spirit, then it is a thing of holiness. How we dwell within it, how we treat it, and how we use it in relationship to other aspects of the planet is extremely important. When we use the body without reverence, we are destructive elements on the planet. We become destructive to ourselves, to other life-forms, and to the earth.

So the issue of whether we see the body as a temple of God or as something we hold with disrespect is an extremely important spiritual issue. Both the Christian cross and the Jewish Star of David are visual symbols of the intersection of the mortal and the divine. In *A Course in Miracles,* in talking about Jesus, it says: "He lived on the earth yet thought the thoughts of Heaven." That is really what we are after from a spiritual perspective. To be very grounded, to live in our bodies in very healthy, vital ways. And to spiritualize the body as well as grounding the spirit.

JOHN ROBBINS: Would you like to close with a prayer?

MARIANNE WILLIAMSON: Oh Spirit, Divine Creator, we pray. We place in your hands our precious earth. We place in your hands all living things. We place in your hands our bodies. We place in your hands our relationship to food. We place in your hands our relationship to animals. We place in your hands our relationship to the earth, and we ask that a great wave of blessing come upon us. May our minds be healed of all illusion and falsehood. May we be lifted in thought and in action. Please purify our thoughts that all behavioral patterns might be lifted to Divine right order, that there is only reverence and love between us and all life. And so it is. Thank you, God. Amen.

JOHN ROBBINS: Amen.

Steps You Can Take: Leaning into Consciousness and Alignment

Write a Food Mission Statement

Write a sentence that describes what you want for your relationship with food, and post it somewhere prominent where you will see it every day.

Give Thanks

Taking a few moments to say grace, to give thanks, or to savor anticipation of a meal can activate your salivary glands and make your whole dining experience feel more nourishing and delicious. It's also healthy for family bonding, and a great habit to expose kids to early and often.

The Great Fridge Swap-Out

Take small steps, one by one, to stock your refrigerator and cupboards with increasingly healthy foods. Start out by looking through your kitchen and choosing one thing that you are going to stop buying. Then the next time you go shopping, put your vision into action by selecting a healthier alternative to replace it.

Make a Food Diary

Expand your food consciousness by tracking everything you eat or drink, including both items and quantities. You can do this for a day, a week, or even longer. This works best if you bring a spirit of self-love and curiosity. After you're all done with your diary, take a look at the data and see what you notice. For a useful template to help you get started, check out the food diary template at: *www.personal-nutrition-guide.com.*

Track Your Mood

Make a chart to track how you are feeling. Do this at a consistent time each day, and give your energy level and happiness level a score in a range of 1 to 10. If you combine your Mood Chart with your food diary, you may get some especially interesting information. If you want to get even more into it, track your exercise level, too.

Find a Great Recipe—and Use It

Find a new healthy recipe that you love, and start preparing it every week or two. Most of us don't have that many things that we prepare on a regular basis, so by adding something good to your "starting rotation," you can lean into a healthier life. When you get comfortable with this recipe, bring in another.

Resources for Feeding Heart, Soul, and Community

Small Planet Fund

www.smallplanetfund.org

Founded by Frances Moore Lappé and her daughter and colleague Anna Lappé, Small Planet Institute is a collaborative network for research and education that brings visibility and financial support to movements working to create citizen-led solutions to hunger, poverty, and environmental devastation around the world.

Marianne Williamson

www.marianne.com

Bestselling author and mentor to millions, Marianne Williamson shares information, inspiration, and tools to help you live a spiritually fulfilling life—and make a difference in the world.

Kathy Freston

www.kathyfreston.com

Kathy Freston is a *New York Times* bestselling author with a concentration on healthy living and conscious eating. Her website supports *The Lean: a 30-day plan for healthy, lasting weight loss*, which includes the plan regimen, recipes, tips for healthier living, and more.

Find Food Freedom

www.darshanaweill.com

Darshana Weill is a Food Freedom Coach. Her website describes how Darshana counsels clients to discover inner awareness and learn tools to support a healthy and conscious relationship with food. Darshana offers Food Freedom courses, free e-books, and inspirational tools.

Local Harvest

www.localharvest.org

Local Harvest is America's number one organic and local food website. The website maintains a public, nationwide directory of small farms, CSAs, farmers' markets, and other local food sources. The search engine can help you find products from family farms, local sources of sustainably grown food, and contact information for small farms in your local area. The online store helps small farms develop markets for some of their products beyond their local area.

Simple Bites

www.simplebites.net/10-tips-for-sustainable-eating

Simple Bites (SB) is a family-oriented online community dedicated to all things food and drink. The website offers a variety of information and resources to support the preparation of real, nourishing food.

Geneen Roth

www.geneenroth.com

Geneen Roth is the author of the number one *New York Times* bestseller *Women Food and God*. She addresses how to turn a painful relationship with food into one of freedom, joy, and peace.

Slow Food USA

www.SlowFoodUSA.org

Slow Food USA is part of a global, grassroots movement with thousands of members in more than 150 countries that links the pleasure of food with a commitment to community and the environment. There are 225 chapters in the United States, in which people focus on local, seasonal, sustainable food; biodiversity (heirloom varieties); small farmers; and food artisans; while promoting the celebration of food as a cornerstone of pleasure, culture, and community.

Community Share Association

www.communityshareassociation.com

Community Share is a private member health association that has created a legally recognized private domain marketplace. This allows members to legally buy and sell to other members their homemade and homegrown foods and medicines, as well as any professional services such as alternative health care, without government licenses, permits, commercial kitchens, or insurance. Annual membership provides members with an online platform for buying and selling.

The RAFT Alliance

www.raftalliance.org

The RAFT Alliance brings together local farmers, chefs, fishers, agricultural historians, ranchers, nurserymen, and conservation activists to make our food system more diverse, democratic, and delicious. RAFT identifies plant and animal foods that are at risk of extinction and works to ensure their survival through seed exchange, education, and mutual support among seed savers, growers, chefs, and farmers.

Vegsource.com

www.vegsource.com

The popular Vegsource website is a comprehensive site and community supporting all things vegetarian. The site features breaking news and insights on vegetarian themes; a resource store; a topic-based discussion forum; recipes; videos from noted and accomplished speakers, doctors, and authors; a lifestyle section; celebrity testimonies; and more.

PART VII

Being a Food Revolutionary

Now we turn the tables, and I get to interview the interviewer, who happens to be my dad, the one who literally wrote the book on *The Food Revolution*.

You may know a little of our family story: My grandfather started Baskin-Robbins, and groomed my dad, John Robbins, to one day succeed him. My dad walked away from what became the world's largest ice cream company—and the money it represented—to follow his own "rocky road," and devote his life to advocating for health and sustainability.

Starting with the million-copy *Diet for a New America*, my dad's books became international bestsellers and helped awaken our society to the possibility of far more compassionate, healthy, and earth-friendly food. He's the recipient of the Albert Schweitzer Prize for Humanitarianism, the Rachel Carson Award, The Peace Abbey's Courage of Conscience Award, Green America's Lifetime Achievement Award, and numerous other accolades. For twenty-five years, John Robbins has been challenging some of the most powerful industries on the planet, and inspiring millions of people to look at their food and life choices as an opportunity to get empowered and make a difference.

What are the most surprising and useful things he's learned from these interviews?

If you want to be a food revolutionary and take advantage of the health secrets we've discovered, what are the top things you need to do?

My dad had polio as a child, and this helped form him into the health advocate he is today. How do suffering and adversity shape us, and how can we seek to use them on behalf of the purposes for which we live?

And how the heck can he bench press his own weight twenty times, at the age of 65?

Of course, my dad's been known to turn the tables and ask me a question or two as well. Here's our lively conversation.

21

Ocean Robbins Interviews John Robbins

OCEAN ROBBINS: What's really at stake today in the world of food?

JOHN ROBBINS: There are powerful forces at work, and they are pushing in very different directions. On the one side there are Monsanto, McDonald's, the agrichemical companies, and the massive vested interests that are behind industrial agriculture, factory farms, and the destruction of family farms. These forces are heavily invested in the status quo. They're profiting from selling food that is harmful to us to eat, and producing it in a way that is harmful to the planet and cruel to farm animals. But they are making a lot of money and are heavily invested in making sure this system continues the way that it is.

On the other side, we have a rising tide of people who want to eat food that is healthy, that has been produced in a way that is friendly to the earth, and that is a celebration of life. These two opposing forces are fighting a war . . . over dinner. I am a journalist reporting on this battle. But I'm also taking a side as an advocate and an activist. I want to help people to answer these challenges in their lives.

Will we let big agribusiness continue to control the entire food supply chain from seeds, to land, to water? Will we let Monsanto and the other biotech companies continue to patent the seeds we depend on for our food crops? Or will we develop food systems that produce nutritious food in harmony with the natural world? Can we put an end to the escalating rates of diabetes, cancer, and obesity? Can we finally make sure that everyone has enough to eat? Can we feed ourselves and our children food that nourishes body, mind, and soul?

OCEAN ROBBINS: We've been exploring some pretty disturbing realities. Genetically modified organisms (GMOs) are spreading all over the world, and are now in 75 percent of the restaurant meals and supermarket items that the average American is consuming. GMOs have been linked to all kinds of problems, including potentially cancer and food allergies. We're learning about dangerous chemicals in our food system, frightening threats to our climate, potential shortages of soil and water, and industries that are profiting from people's ill health. A lot of us feel overwhelmed just trying to cope with our lives as is, and making a change or trying to face so much destruction can feel like too much. Do you ever feel like you're making people's lives more difficult by sharing so much depressing information? Is ignorance sometimes a form of bliss?

JOHN ROBBINS: I actually don't think, in matters like these, that ignorance is bliss. I think ignorance is accommodation; ignorance is supplication and subordination. When we choose to remain ignorant, we acquiesce to our own oppression. The most powerful weapon any oppressor can wield is power over the mind of the oppressed.

Ignorance can seem blissful, because sometimes the process of awakening, of becoming aware of things that are disturbing, is painful. But that pain is the breaking of the shell that otherwise encloses our understanding, our capacity to act, and our ability to make a difference.

If we remain encased in ignorance, we don't do anything differently and we in effect condone, and become partners in, the destruction that's taking place. When we become conscious, then we start to be able to actually make choices that make a difference.

Our way of eating is deeply intertwined with an economic system, and with a belief system about the human relationship with the natural world. In many cases, we humans have been exploiters. We've preyed on the earth and its creatures, and even on one another. In the process we've done a tremendous amount of damage. There are so many of us now on the planet that if we continue to cause as much damage per person as we have been, the consequence will be our own demise.

We're doing something that's utterly out of phase with our own greatness, our own beauty, and our own love. We are literally destroying the earth on which we depend for our food, for our economy, and for our lives. The effort to bring our food systems, our personal choices, our ways of life, and our definition of success back into alignment with the well-being of the whole is a tremendous challenge. What I hope is that in bringing this information to people and raising these questions, we can enliven people's ability to experience themselves as positive and creative vehicles for change in their lives and in the world.

OCEAN ROBBINS: What are the benefits that we can look forward to as we participate in the food revolution?

JOHN ROBBINS: Greater health, greater vitality, less disease, a stronger immune system, a stronger respiratory system, a stronger cardiovascular system, a longer life, more joy, more sensuality, more pleasure, and a greater capacity to respond creatively to life's challenges with resilience and resourcefulness rather than with resignation and passivity.

OCEAN ROBBINS: You've been a passionate gardener for much of my life. And every Saturday morning you grab a bunch of canvas shopping bags and head over to the local farmers' market. Why do you love growing food and shopping at farmers' markets?

JOHN ROBBINS: When you grow food on any scale, it helps you to experience your connection with the earth, and to take pleasure in the earth's fertility. Second best to growing your own food is to buy directly from the grower. Farmers' markets are places where you can meet the farmers and buy directly from them.

When you buy from a grower, there's a direct contact in that—to the farmer, and even to the land. For those who have access to one, a farmers' market is a great place to get food that is fresh, nutritious, and beautiful.

OCEAN ROBBINS: I think food should be fun. It should be a source of beauty and happiness. When we nourish our bodies, we are truly making a statement to ourselves and to our world that we want to care for ourselves and that we are participants in a larger community of life. There is something extraordinary about stepping lovingly into the web of relations that consuming food represents. Bringing joy to our relationship with food adds a bounce to the step of our lives. There is a lot of heaviness in some of the data, but when we can actually embody what is possible, we can find tremendous richness and beauty.

JOHN ROBBINS: And pleasure. Like you and many others, I love food. My diet is relatively simple. I don't eat very many processed foods, and I don't eat a lot of fat. Some people would consider my diet Spartan. But the truth is, I get more pleasure now out of eating say a simple baked potato than I used to out of eating a baked potato smothered in sour cream and butter and beef gravy—because I am more present. I feel more centered in my body; and I feel more connected to my senses.

When you're healthy and you're living a life that's true to your essential spirit, then the simple pleasures of life are fulfilling. I believe that eating simple food in a healthy body with a clean conscience is more pleasurable, and infinitely more satisfying, then eating decadent food that makes you and your world ill.

OCEAN ROBBINS: Healthy food is an important facet of fitness, but you're also an advocate for exercise. And you really do walk your talk. When I was about 8 years old, you introduced me to running, and became my coach. By the time I was 10 we were running marathons together, and you were competing in triathlons. Now we work out at the gym together, and even though you are a 65-year-old grandfather, you still regularly bench press your weight twenty times. I don't know if many people of any age can do that.

Your accomplishments are exceptional, but your message has always been accessible. Wherever we are in the fitness spectrum, we can all do something. Whether we get a little exercise or a lot of exercise, we can lean into the health that we want to have. Over time, our body can become a source of increasing fulfillment and joy. What does exercise mean to you, and why do you think it's so important?

JOHN ROBBINS: When I was 6 years old, I had a form of polio and it left me for a time in a wheelchair and unable to move my legs. Eventually the feeling and movement returned to my legs, but it did so much less in my left leg than in my right. When I was 21 my left leg was three inches shorter than my right leg and was very weak; I could hardly put weight on it and it was fairly withered. The doctors at the time told me, "John, your growing years are over. You're just going to have to live with it." Have you ever noticed that doctors don't always know everything?

OCEAN ROBBINS: We sometimes act as if M.D. stood for Medical Deity.

JOHN ROBBINS: Yes, and then we supplicate before anyone in a white coat. But I didn't accept the prognosis I was given. Instead, I changed my life. I stopped trying to live up to the agenda that my father had for me. I changed the way I ate completely—from an ice cream–centered to a plant-strong diet; from a lot of processed and artificial foods to natural foods. I began to do yoga, and I began to meditate. I began to do a lot of deep therapy and work on myself to investigate and explore my own beliefs and my own assumptions so that I could work with them. What in fact happened was that over the next ten years, my leg did grow, becoming as long as the right one. It never got fully as strong, but now it comes close. I always love the opportunity to hike, to climb mountains, and to be athletic. I never take the ability to do these things for granted. I am grateful beyond description that my body works as well as it does, and I want to make the most of it.

Our capacity to respond to the suffering and pain in our lives is very important. Sometimes we find that we have the ability to activate rather than merely to accommodate. We learn how to use our suffering for growth, for learning, to deepen our empathy, and to make us stronger at the broken places. For me, exercise has been a way of expressing a commitment to living fully, and to engaging with suffering and stress in a positive way. Every time I work out at the gym, every time I run a race or just jog or hike, I feel physically alive. I'm doing it as an expression of gratitude for the opportunity to live in a body that works.

I'm grateful that I can breathe. I'm grateful that I can walk. I'm grateful that I can do these things that I enjoy so much and that are a beautiful part of being human.

We can take our health for granted. When I was in high school and I was, to a large extent, unable to express myself physically, I used to admire the football stars and the track team captains, who had natural athletic gifts. I know some of these guys today, and a lot of them have heart disease and are overweight. They haven't taken care of themselves because they took their gifts for granted. I didn't take them for granted, because I didn't have them. I'm actually now grateful for the fact that I struggled so much physically, because it spurred me to do what I could with what I had. I think when we use what we have to the fullest extent possible, we get more. We get more responsibility, and we get more ability.

OCEAN ROBBINS: If you want to know how important breathing is, you can just talk to someone who's had asthma, lung cancer, or emphysema.

JOHN ROBBINS: When I see people smoke, I know the emptiness that they are probably dealing with and that there is pain and loneliness inside them. I hope they are getting some relief, but I also know that the habit is a dangerous one. We need to support each other with love, kindness, and compassion so that we can shift out of the destructive habits and create natural highs. We need to develop ways of getting our needs met, and of finding our relief and our pleasures, that are consistent with our greatness as human beings rather than our destruction.

OCEAN ROBBINS: Only about 25 percent of the 141 medical schools in the United States require their students to take a single nutrition course over all their years of training. Why are these institutions not educating future physicians about the foundation of health?

JOHN ROBBINS: The practice of medicine as we know it is not actually designed to support the health of people.

It's really designed to diagnose and treat disease—primarily with drugs and surgery. We might think that a doctor would be someone who would know about preventing disease and promoting health. But in fact, compared to others, our medical doctors live shorter lives. They have a higher rate of drug addiction, and they have more heart disease. They aren't examples of health, because they haven't learned the principles of healthy living. Their education has taught them a great deal about drugs, surgery, treatments, and procedures. But they've been taught little to nothing about the practices and principles that actually prevent disease in the first place. Doctors are working in a system that pushes them really hard, and they are usually so overworked that they don't have time to investigate on their own. They haven't been taught, and generally don't learn, much if anything about prevention.

Fortunately, there are many health researchers all over the world who have taken it upon themselves to study prevention and the role of nutrition. Some of them are M.D.s like Dr. Esselstyn, Dr. Ornish, Dr. Fuhrman, and Dr. Barnard. Because of their work, we now know how to prevent many of the diseases that afflict us today. We all owe a huge debt of gratitude to them for their courageous work.

OCEAN ROBBINS: With all that we know about diet and health, why do you think so many people continue to eat food that is killing us? And perhaps more importantly, what can we do about it?

JOHN ROBBINS: Many of us are drawn to eat unhealthfully because processed junk foods are tasty, convenient, and inexpensive. Part of the reason we find them tasty is that they are familiar. They are convenient because they are widely available and heavily marketed. And they are inexpensive in part because they are subsidized.

It's crazy, but government policy actually subsidizes those foods that are least healthy. We subsidize genetically engineered corn and soy, which become livestock feed in factory farms, high-fructose corn syrup, isolated proteins, and hydrogenated vegetable oils. Why don't we instead subsidize healthy foods? What if we were to tax white bread and then use the revenue to lower the cost to the consumer of whole wheat bread? What if we were to tax pesticides and use the income from that tax to lower the cost to the consumer of organic food? What if we taxed sugary sodas, and used the income from that to lower the cost to the consumer of fresh fruits and vegetables? There are revenue-neutral actions we can take to not just level the playing field, but also to tilt it in the direction of greater health.

OCEAN ROBBINS: There are about 46 million Americans that use the food stamp program to buy food. People in communities that depend on the food stamp program tend to eat less fruits and vegetables and more highly processed, heavily packaged, and tragically unhealthy foods. It's little wonder, then, that some of the poorest communities in the United States also have the highest rates of cancer, diabetes, heart disease, and all the other preventable illnesses that plague the modern world. Do you think there's anything that could or should be done with the hundreds of billions of food stamp dollars to, as you say, "tilt the playing field" in the direction of greater health?

JOHN ROBBINS: At present, food stamps cannot be used for tobacco or alcohol purchases, but it is true that they are frequently used to purchase foods that we know will, in the long run, make people sick. What if we were to restrict food stamps to be used for real and healthy foods? If we did that, then consumer demand would necessitate that the convenience stores, 7-Elevens, and other food sources that are often the only options available in lower income communities would start to provide fresh fruits and vegetables. Changing what foods stamps are used for would go a long way towards helping lower income people to live healthier lives. This would lower medical costs and lower the rates of obesity and diabetes. It would help millions of people to feel better, and to be more capable of raising healthy children and producing healthier communities. In the long run, the health-care savings would be tremendous, and this would help the whole economy.

OCEAN ROBBINS: If food stamps were used only for healthy foods, then I don't know that ice cream would be on the list. But your dad, Irv Robbins, and your uncle, Burt Baskin, started a multibillion-dollar ice cream company. When you were in your teen years, and your dad was grooming you to one day run the company, the advertising department came up with a new slogan. It was: "We make people happy." Your dad was quite excited about this slogan and it did turn out to be extremely successful in terms of selling ice cream. But you were disturbed by it.

JOHN ROBBINS: Baskin-Robbins based its whole ad campaign for a year around that phrase. I told my dad at the time that I didn't like it. He loved it and of course he had the final word. But I said, "I don't like it because I don't think it's true. We don't make people happy. We sell ice cream, which is a product that provides momentary pleasure. Human happiness is a very difficult thing. It's challenging. It's certainly not something you can buy." My dad responded by saying, "We're just trying to sell ice cream here. Stop with the philosophy." I said, "That's the point. We're selling ice cream; that's what we do. We don't make people happy, we sell ice cream."

My dad was not pleased with my way of thinking, and I admit I wasn't especially pleased with his either. What was becoming apparent to me was that he and I weren't on the same page at all. We didn't really have the same agenda. It was very painful for me to realize it, but I was being called in a different direction than the one he wanted for me. He really did want me to one day join him in running the company, but I was coming to realize that I had a different path.

Not long after our conversation about the ad campaign, my uncle Burt Baskin died of a heart attack at the age of 54. He was a very big man who ate a lot of ice cream. I asked my dad when that happened if he thought there might be a connection between the amount of ice cream my uncle would eat and his fatal heart attack. My father looked at me and he said, "Don't ever ask that again." Then he said, "No. His ticker just got tired and stopped working." I realized that my dad simply could not bear to consider the possibility that there might be a connection. And I understood why. By that time he had manufactured and sold more ice cream than any human being who had ever lived on the planet. He didn't want to think that the family product was harming anybody, much less that it could have played a role in the death of his beloved partner and brother-in-law—my uncle. But even if he couldn't consider the possibility, I felt I should. I didn't want to be involved in selling a product that might be undermining people's health.

The problems with ice cream are considerable. Eaten in excess, it can cause great harm. Ben Cohen is the founder and was for many years the co-owner of Ben & Jerry's. Ben was forced to have a quintuple bypass procedure performed on his heart in his late forties. Ben is a very fine man who has done a lot of good in his life. He is also a big man who ate a lot of ice cream. I am not saying that an ice cream cone is going to hurt anybody. But the more you eat, the more likely you are to have heart disease, the more likely you are to become obese, and the more likely you are to become diabetic.

I decided that I didn't want to make my living and source myself in the sale of a product that might harm people. I made a decision to walk away from the company. And to be ethically consistent, I told my dad that I didn't want to have a trust fund. I didn't want to depend on his financial achievements for my life, and I walked away from the money.

OCEAN ROBBINS: The media called you "The prophet of nonprofit."

JOHN ROBBINS: They called me a lot of things. What was it like for you, Ocean? You grew up knowing my dad—your grandfather—and of course eating a very different diet than I did as a child. What was your experience of that process?

OCEAN ROBBINS: I grew up in a family that grew a lot of its own food and loved eating simply. You used to quote Henry David Thoreau: "I make myself rich by making my wants few." My early childhood was infused with those kinds of values. I think it's actually hard for you and me to grasp what different childhoods we knew. You grew up in a family with a father who was working maybe seventy hours a week, building a huge company, and selling ice cream. I grew up in a family with parents who did yoga and meditated, and called their son "Ocean." In the home I grew up in, love and simplicity were more important that worldly accomplishments. Our lives had very different beginnings. And yet, remarkably enough, we have highly resonant values. I feel blessed that in my own life I've been able to use your life, your values, and all that you've learned as a point to grow from. I feel that I grow out of rich soil.

I remember in my teen years when a lot of my peers were rebelling against their parents. I decided that there actually wasn't much point in rebelling against my parents, because to do so would also mean rebelling against my own values.

It was more interesting to me to rebel against poverty, violence, and environmental destruction than to rebel against
my family. I needed to differentiate; I needed to know
myself as an individual. I deeply wanted to find my own
unique contribution to the work of our times. You have
really been a friend and an ally in that quest throughout my
life. And now I'm so grateful that we have the opportunity
to work together.

JOHN ROBBINS: As these interviews have unfolded, what
have you learned? What has been meaningful to you?

OCEAN ROBBINS: I've been inspired by the consistency
with which the conversations have pointed in some very
common directions. Each has its own perspective and its
own contribution to make to the whole picture. But when
you look at it together, it's clear that we need to make some
big changes. I am also struck by the importance of holding
our journey with compassion.

If we set an impossible standard for ourselves, then we
are going to fail, and then that failure is going to only add
to the sense of misery on this planet. But if we take one step
at a time, if we lean into what's possible, then we really can
be a part of something beautiful. Every single step we take
matters. You don't have to go 1,000 miles to take a step. I
think moving in the right direction is profound. And with
every step you take, the next one gets easier.

As bad as things are in this world, as much suffering and
violence as we experience, that's how much better things
can be with a change. Just imagine: what would it be like to
have really healthy arteries carrying healthy blood through
your whole body? What would it be like to live in a world
where more and more and more people had enough to eat,
where our farms were healthy and life-giving, where there
was more topsoil every year, where our water systems were

clean and sufficient, where a stable climate was something we could count on? All of that is completely possible.

What you do matters, and it matters a lot. And yet, it's not enough just for individuals who can afford it to start making choices to go organic, eat locally, and bring more fruits and vegetables into their lives. We need to look at the policies and politics of our food systems.

When you look at the political landscape, what do you think are some of our key leverage points—the spots where we could make a huge and positive change on a systemic level?

JOHN ROBBINS: I see several key places where we can make a substantial and immediate impact.

1. **Label genetically engineered foods.** If you don't want to eat genetically engineered foods, you should have the information that lets you make a choice that is consistent with your desires. But there's another reason you should have a right to know.

Right now if a mother feeds her baby Gerber's baby food and it's made with cornstarch, that cornstarch is probably made from genetically engineered corn. If the baby gets sick, the mother has no way of knowing if the genetically engineered corn that was used to make the cornstarch that's in the baby food might have been the cause. If the labeling were there and that happened, she might go to her pediatrician and say, "I don't know if there's a connection, but my baby got sick when she ate this genetically engineered food." The doctor then, assuming there was a reporting mechanism, could report that information and gradually the Centers for Disease Control and other groups could be able to do some surveillance and be able to track what was going on.

Right now we're part of an enormous experiment—we're all guinea pigs. But no one is actually paying attention to the results. There is no tracking. We have no way of tracing these foods and the problems they might be causing.

What tests have been done are very troubling. We see toxicity, liver damage, and digestive disturbances in every type of lab animal that's been fed genetically modified foods. But the big test that's being done, the one in which we are the guinea pigs, isn't being monitored. I think labeling is a tipping point. We need to get labeling.

2. Move away from factory farms and back towards family farming. The whole industrial model of meat production is only cheap because it externalizes the costs onto society. If factory farms and feedlots had to pay full market value for the feed crops and water they use, the price of the products would be so high that Americans would have to treat them less like a staple and more like a garnish. If factory farms had to pay their own pollution costs instead of getting taxpayers to both pick up the tab and also suffer the consequences of the pollution, the whole model would no longer be economically viable. We need to stop subsidizing this tragic way of producing our food. And we need to stop the routine use of antibiotics in factory farms, so that we cease breeding antibiotic resistant bacteria, so that our antibiotics can retain their capacity to be effective medicines into the future.

3. Get over our schizophrenic attitude about animals. On the one hand, most people love animals. We treat our pets very well. We feed them. We pay for their food. We pay for their vet bills. They often sleep on our beds. We consider them part of our families. We lavish our affection on them, and we get a great deal in return. I think the relationships that people have with their companion animals are often extremely enriching to us as human beings, and that is beautiful.

On the other hand, though, we call other animals "livestock," or food animals. And by virtue of making that semantic distinction, we feel entitled to treat those animals with any manner of cruelty, as long as it lowers the price per pound.

If you are a meat producer or a dairy producer, you can treat your animals with any manner of cruelty as long as the practices you are employing are considered "customary." As long as it's considered standard farming practice, the degree of suffering that the animals endure is not a factor in what is legally permissible.

More and more people are lifting the veil and asking real questions. How has this food been produced? People don't want their chocolate to be harvested by child slaves. They don't want their coffee beans to have been grown by people laboring in sweatshop conditions. They don't want to buy tomatoes that have been harvested by people paid wages so low that they are in effect enslaved. And they don't want meat and dairy products that come from animals raised in conditions akin to concentration camps.

OCEAN ROBBINS: More and more people care, and are seeking out products that are in alignment with their values. But some industries try to take unfair advantage of consumer's ethical concerns with deceptive marketing. You actually joined in a lawsuit against leaders in the California dairy industry for their Happy Cows ad campaign. Why do you find their ads so disturbing?

JOHN ROBBINS: The California dairy industry has spent hundreds of millions of dollars trying to convince us that California cows are happy. But there's no way that the life of the average California dairy cow today could legitimately be considered a happy one. The natural life span of a dairy cow is about 20 years, but in California dairy cows are usually slaughtered by the age of 4 or 5. By that time, they're typically crippled from painful foot infections and from calcium depletion.

The ads show happy cows grazing in lush, green fields, with slogans like "So much grass, so little time." But some of these ads were actually filmed in New Zealand. Most of California's dairy cows are confined in dry feedlots in the barren San Joaquin Valley, where they may never see a blade of grass in their lives. There are often 15,000 or 20,000 cows in one of these dairies. These are not family farms. The ads present the California Dairy Industry as a bucolic enterprise that operates in lush grassy pastures. But that's not the truth. In reality, California's dairies are more confined and more inhumane than the national norm.

If someone is willing to pay extra for humanely raised animal products, they're expressing a kind of compassion that I think is admirable. But when an industry says that their dairy products are from happy cows when they are not . . . to me that is deceitful. It's false advertising, and it's exploiting the very people who are concerned, who are willing to put their money where their mouth is, and who really want to help the world be a better place.

The Milk Board knows that showing the actual reality—showing calves being ripped away from their mothers and being confined in tiny veal crates—would not sell their product. They know that if they showed the reality—emaciated, lame animals who have collapsed from a lifetime of hardship and overmilking, animals being taken to slaughterhouses and having their throats slit—this would not sell their product. But this is the sad reality of the California Dairy Industry. Covering up this misery with fantasy ads of happy cows does absolutely nothing to alleviate the real suffering that these real animals endure.

OCEAN ROBBINS: *Diet for a New America* was first published in 1987. I was 13 at the time. In the five years after its publication, beef consumption in the United States dropped by nearly 20 percent and the meat industry was not exactly pleased with you. They spent a lot of money actually trying to discredit you and to destroy your work. They were paying academic institutions to write reports that sought to disparage your conclusions. They were calling on ranchers to mail you "gifts" of manure. They were shadowing your tours, trying to disrupt your public talks and your media appearances. They were making threats, some of which were pretty disturbing. Were you ever frightened by all of this, and what kept you going?

JOHN ROBBINS: I was frightened at times by the death threats and to tell you the truth what frightened me the most was when your life was threatened. But ironically what kept me going was you. I would look at you and I would see so much innocence and hope. Seeing your beautiful soul would remind me that all children deserved to live in a world with clean air, clean water, and a viable economy. That everyone deserves a stable climate and a food system that produces food that's nutritious, healthy, and supports us in being the kinds of human beings we want to be.

So when I would receive these threats, I was, at times, frightened, but it actually made me more determined. If they had to resort to that level, then I felt there was all the more reason to bring about a revolution in our way of thinking about our food, our way of producing it, and our way of eating it. I actually think that in many ways I was strengthened by the attacks.

OCEAN ROBBINS: Sometimes the greatest challenges we face can become something that we find a way to use on behalf of what we love. I think that growing up in an ice cream family taught you something not just about the madness of the food industry, but also about how important it is to create something different. What would you tell our readers who are inspired to make a change, but don't know where to start?

JOHN ROBBINS: You start exactly where you are. If you quiet your mind and open your heart, you'll know what to do. I think you do it a step at a time, a day at a time, a mouthful at a time, a purchase at a time, a meal at a time. And I think we will change the world, one heart at a time. The primary agent of social change, to my way of seeing, is the human heart. As it awakens, as it becomes engaged, as it becomes alive with its responsibility, with its capacity for wisdom, for caring, and for soulful action, then our prayers start to manifest in our lives. Then our caring starts to be expressed in the way we are with each other. Then we start to bring out the best in one another. Then we live in a way that's worthy of the suffering we have endured, and that begins to transform the suffering into strength.

∼ **Being a Food Revolutionary** ∼

Thank you for all the ways in which you participate in the food revolution. One farm, family, and choice at a time, this is a movement that is changing the course of history. It's a movement that grows stronger . . .

Every time you choose spinach over Snickers.

Every time you choose organic and locally grown over chemical-laden, genetically engineered pseudofoods.

Every time you choose to set a positive example instead of succumbing to the status quo.

Every time you make a nourishing meal and enjoy it with reverence.

Every time you support healthier food systems by signing a petition or lobbying a representative or organizing in your community.

Every time you speak out instead of being dumbed down.

And every time you look at food as an opportunity to take a stand for the life you want and the world you love.

We wish you radiant health, abundant joy, and a life filled with meaning. And we wish you good, delicious, nourishing food.

—John and Ocean Robbins

~~~~ **Will You Join Us?** ~~~~

If you want to be part of the movement to bring forth healthy, sustainable, humane, and conscious food for all, we're here to help you take the next step.

For extremely current information, inspiration, and empowerment to help you be a food revolutionary, connect with the authors at *www.foodrevolution.org*.

You will find tools, leading-edge information, summits with extraordinary speakers, online courses, inspiring articles, a food revolutionary bookstore, and a comprehensive resource guide to help you live and share the message of this book.

www.foodrevolution.org

～～～ **About the Authors** ～～～

John Robbins is the author of the bestselling *The Food Revolution*, *Diet for a New America*, and *No Happy Cows*. His life and work have been featured on PBS.

Ocean Robbins founded and for 20 years directed Youth for Environmental Sanity (YES!), starting at age 16, and is now CEO of The Food Revolution Network.

They both live in the Santa Cruz Mountains. Visit them at *www.foodrevolution.org*.

To Our Readers

Conari Press, an imprint of Red Wheel/Weiser, publishes books on topics ranging from spirituality, personal growth, and relationships to women's issues, parenting, and social issues. Our mission is to publish quality books that will make a difference in people's lives—how we feel about ourselves and how we relate to one another. We value integrity, compassion, and receptivity, both in the books we publish and in the way we do business.

Our readers are our most important resource, and we appreciate your input, suggestions, and ideas about what you would like to see published.

Visit our website at *www.redwheelweiser.com* to learn about our upcoming books and free downloads, and be sure to go to *www.redwheelweiser.com/newsletter* to sign up for newsletters and exclusive offers.

You can also contact us at *info@rwwbooks.com*.